IMAGINARY

CARTOGRAPHIES

Imaginary Cartographies

Possession and Identity in
Late Medieval Marseille

Daniel Lord Smail

Cornell University Press

ITHACA AND LONDON

First published 2000 by Cornell University Press

Printed in the United States of America

Library of Congress Cataloging-in-Publication Data

Smail, Daniel Lord.
Imaginary cartographies: possession and identity in late
medieval Marseille / Daniel Lord Smail.
p.   cm.
Includes bibliographical references and index.
ISBN 0-8014-3626-5 (cloth)
1. Cartography—France—Marseille—History.   2. Geography,
Medieval—France—Marseille Maps.   3. Marseille (France) Maps.
I. Title.
GA865.M37S63   2000
526′.0944′9120902—dc21   99-41668

Cornell University Press strives to use environmentally responsible suppliers and materials
to the fullest extent possible in the publishing of its books. Such materials include
vegetable-based, low-VOC inks and acid-free papers that are recycled, totally
chlorine-free, or partly composed of nonwood fibers. Books that bear the logo
of the FSC (Forest Stewardship Council) use paper taken from forests that have
been inspected and certified as meeting the highest standards for environmental
and social responsibility. For further information, visit our website at
www.cornellpress.cornell.edu.

Cloth printing   10   9   8   7   6   5   4   3   2   1

Contents

Illustrations

Tables

Preface

This book is a study of imaginary cartographies in the city of Marseille in the later middle ages. The study of urban cartography in this time period, roughly between 1300 and 1500, is invariably a study of imagination and language, because late medieval cities were rarely represented in their own time by means of graphic maps or other visual images representing a city from a perpendicular viewpoint. Yet there was a map— or maps, as I shall argue throughout this book—informed both by cartographic lexicons and cartographic grammars. A cartographic lexicon consists of all the toponyms or place names that speakers of a shared language attach to their landscape. These languages, in turn, configure toponyms according to a cartographic grammar, a linguistic or cognitive framework that I shall call a template. Together, toponym and template constitute a cartographic science, or a way of knowing and classifying space.

Today the technologies of surveying and the address templates particular to given countries have combined with the institutional authority of the state to create a universalizing cartographic lexicon and a common grammar of plats, streets, cities or towns, and ever-larger geo-political entities. This embraces not only the cartographic science of plats or cadasters but also conventional political atlases and city or local maps. If maps can also be used to represent much more, such as physical topography or flora, these are seen as thematic maps, a category somehow apart from the norm. There was no such universal cartography in late medieval Marseille, although some universalizing forces were at work in the cartographic practices of public notaries and seigneurial officials. It is the multitude of available grammars and the presence of two competing languages, the Latin of record-keeping and the vernacular languages of given cities and regions in Europe, that make the subject and the time period so especially important.

I will be arguing throughout this book that slight differences in grammatical usages and lexical terms found in the sources from medieval Marseille reflect major differences in cartographic imagination between various linguistic communities. These differences, in turn, reflect how cartography and space can be manipulated in the play of power and interest in a late medieval city. To a certain extent, therefore, I have taken up Peter Burke's call to study language as a social institution, and in what follows I will be exploring how historians might use recent theories in sociolinguistics and the sociology of language. For my purposes, the ability to draw meaningful conclusions from a study of sociolinguistic variations makes it possible for historians to explore language serially and statistically. I have made words and language the subject of a social science methodology.

This book is also a history of the record-keeping bureaucracies and practices that were in the process of refining this cartographic science. Record-keeping bureaucracies, as I have used the expression here, are the institutions or the spheres of activity within institutions that keep track of people, landed property, and other possessions, and in so doing necessarily create, shape, classify, and record conventions or configurations of identity. As a general rule, the classificatory schemes used by medieval agents of record did not depart significantly from what one might call folk classification, the ordinary schemes of classification typical of spoken language. Bureaucratic identity clauses did not clearly distinguish an entity—an individual or a possession—from the social context in which that entity was known by others. One did not need a lengthy abstract description of identity because the agent of record, like other interested parties, simply knew the person or possession in question. Even in the later middle ages agents of record often used verbal descriptions, if they used them at all, to jog their memory. As a result, identity clauses did not necessarily attempt to signify unique people or entities in mechanical or rational ways. The proof of this lies in the absence of any conventional identity formulas in the records from Marseille and the paucity of available identity categories. Even so, there is some evidence that this situation was beginning to change.

Today, the record-keeping activity of state bureaucracies is a product of an epistemological transformation in the ways in which bureaucracies know individuals. Modern rational-legal bureaucracies track the identities of people and possessions by means of precise templates that provide an abstract and impersonal description of a unique individual or a unique property. The frequent use of numbers in these templates, such as the numbers used in identification cards or those found in street addresses and postal codes, is but one expression of the mechanical rationality of these templates. Such templates are the product of a modern science of bureaucratic classification, and depend upon the historical invention of multiple and precise categories of identity or, in the case of landed property, on the sophisticated and expert technologies used in surveying and map-making. Here, I will be concerned with schemes of classification that were just beginning to emerge in record-keeping practices in the later middle ages. How did notarial identity clauses signify men, women, and property? Why was a bureaucratic science of abstract classification emerging? The developing identity templates reflected common usage, surely, but did they in turn *shape* common usage? Did they change the ways in which people perceived themselves, mapped their possessions, and located themselves in the world? There are good reasons to think so, given the sheer scale of the contact between growing record-keeping bureaucracies and the general public. But if the emerging templates did have this effect, what does this mean for the social histories based on this changing bureaucratic science of classification?

This is not the place to answer all these questions, because the subject, properly framed, spans continents and centuries. My book is restricted to a period of time stretching from the late thirteenth to the mid-sixteenth centuries, and the bulk of the material comes from the mid-fourteenth century. I have used only the archives of the city of Marseille because I need a sufficient density of documentation to do many of the things I do in this book, such as trace and compare the various addresses and physical locations given by or assigned to a large number of men and women and units of property from the period in question. But I am conscious of the need to provide comparison. Friends and colleagues have been more than patient with my endless and seemingly irrelevant questions about the plats and addresses found in the records from medieval or Renaissance Crete, Venice, Rome, Florence, Montpellier, and Perpignan, and I have learned things that both corroborate and challenge my larger thesis. To the best of my knowledge, however, little of this information has been published in forms that I can readily compare to my own data, and since the information remains largely impressionistic I have made a decision not to incorporate it into this project. My hope is that this book will stimulate the research that will make larger syntheses possible. I expect that many of my conclusions will be undermined or nuanced when the field of study is extended beyond Marseille, but I will be gratified if future work upholds two of my more general arguments: (1) that many late medieval record-keeping bureaucracies were something other than institutions of state and governance, and that the early evolution of the identity science used by record-keeping bureaucracies was not exclusively or even primarily a state-driven process, and (2) that the geographic component of the identity templates developing in the later middle ages was shaped not by the political interests of states but rather by the culture of a relatively small group of late medieval agents of record and, in particular, public notaries.

During the course of writing this book I have acquired numerous debts of gratitude. My dissertation adviser at the University of Michigan, Diane Owen Hughes, has been a never-failing source of inspiration; from her I have learned that there is always something more interesting to be said about the things one reads and observes. I cannot thank Thomas N. Tentler enough for his constant presence. I also owe a great deal to Thomas Andrew Green, Raymond Grew, Sally Humphreys, William Ian Miller, and David Bien. I first began wondering about imaginary maps in conversations and exchanges with Sujata Bhatt, and I thank her for ideas upon which I have drawn extensively. I received generous funding from the Horace H. Rackham School of Graduate Studies at Michigan, and the research I did in 1990–1991 on which this book is based was funded by grants from the Lurcy foundation and the Mellon foundation; I am grateful to both for their generous support. A grant from Fordham University allowed me to return to the archives in the summer of 1996 for additional work, and I thank the Fondation Paul-Albert Février for the privilege of residing in the Février apartment during my visit. A fellowship awarded by the American Council of Learned Societies for research into my next project allowed me to take a sabbatical leave in the spring of 1998, and it was during the interstitial moments of archival research that I completed revisions to the final draft of this manuscript.

In France, I owe thanks to M. Noël Coulet and to Mme Marie-France Coulet, who welcomed my wife and me and our children to Provence and housed us in a villa of unforgettable beauty in downtown Aix-en-Provence. M. Coulet provided much guidance to the archival wealth found in Marseille and unfailingly answered my many and diverse questions over a period of nine months, as did M. Louis Stouff. Mme Claire Laurent at the library of the Centre des Études des Sociétés Méditerranéennes was unfailingly helpful and directed my use of the marvelous collection she oversees there. I owe thanks to M. Pierre Santoni and especially to

M. Félix Laffé and M. Alain Lassus at the Archives Départementales des Bouches-du-Rhône, to the former for his care in introducing me to archival material and for helping me to read Provençal, and to the latter for his generous assistance in helping me microfilm much of the material I needed for my research. The Archives Municipales de la Ville de Marseille, in turn, is surely one of the most agreeable environments in which anyone ever worked, for which Mme Isabelle Bonnot and now Mme Sylvie Claire and their gracious staff are responsible. It was a pleasure working there and a pleasure to acknowledge all their help. M. Marc Bouiron, staff archeologist for the city of Marseille, has shared with me his profound knowledge of the physical and social contours of the medieval city and also prepared maps in chapters one and four; I am very grateful for his generous efforts on my behalf.

Carol L. Lansing and L. R. Poos read parts of the dissertation from which this book emerged and provided collegial advice on this and other projects. Kathryn L. Reyerson has been involved in this project from the start and on numerous occasions has given me the benefit of her expertise on southern French social and urban history. Thomas Kuehn has been a tireless reader and critic of my ill-formed thoughts and has been a powerful influence on my understanding of the legal issues raised in this book; my debt to him is great. Charles Burroughs has provided me with helpful bibliographic suggestions regarding the Italian peninsula, and Mme Claude Gauvard has done the same for recent bibliography concerning France. At Fordham University I have learned from conversations and exchanges with Thelma Fenster, Richard Gyug, Joel Herschman, David Myers, Tip Ragan, and Susan Wabuda. Maryanne Kowaleski, Kathryn L. Reyerson, Teofilo Ruiz, and Orest Ranum all went through a first draft of the manuscript in painstaking detail, and Thomas Kuehn read a revised draft. I benefited enormously from their numerous suggestions and reflections; theirs, collectively, is the very model of scholarly criticism. Portions of this project have been presented at various venues of the last few years, and I would like to thank those who have organized these events: Kathryn L. Reyerson and John Drendel, H. Wayne Storey, and Barbara Hanawalt and Michal Kobialka. I have also benefited from conversations and correspondence with Benedict Anderson, Gail Bossenga, Molly Greene, Michael Herzfeld, Lynn Hunt, Julius Kirshner, Derek Keene, Megan Koreman, Mary Lewis, Sally McKee, Edward Muir, Ann Mulkern, Shlomo H. Pick, Jacques Revel, Raymond Van Dam, Laure Verdon, Lauri Wilson, and Rebecca Winer. If I have not made the best use of their advice that is my own fault; it goes without saying that I am responsible for the flaws that remain. Finally, I have been inspired by the enthusiasm and encouragement of my editor at Cornell, John Ackerman, and would like to thank him for all he has done for this book.

My mother, Laura L. Smail, has added a measure of grace and consistency to the writing through her careful editing, and it saddens me that my father, John R. W. Smail, is no longer able to bring to this the wit and insight he has brought to the work of so many of his own students. Both my parents contributed to this in ways that go beyond my power to acknowledge with any sense of adequacy. My brother John offered useful advice on computers and databases that made the research for this book go smoothly. My wife Kathleen and I have built a life of humor and ironic detachment that takes in stride our disordered lives; it is an environment in which our children, Benedict, Irene, and Gregory, thrive. It is to them that this book is dedicated.

DANIEL LORD SMAIL

New York

A Note on Names

In the fourteenth century, Marseille was part of the county of Provence, itself attached to the Kingdom of Naples, and the everyday language of the city's residents was Provençal, into which intruded certain words of Latin or French origin. The French language can be found as a language of record in Marseille as early as 1465, before the city itself fell into the realm of the French kings, but in the fourteenth century the residents of Marseille considered French a foreign language. Latin was the most common language of record; this, too, was not the language of the vast majority of the inhabitants of Marseille. In an attempt to be faithful to the Provençal culture of fourteenth-century Marseille I have generally used Provençal names both for cartographic sites and for individual names. Thus, a man known as *Johannes* in Latin records and as *Jean* in modern French was known as *Johan* in Marseille; it is *Johan* that I have used throughout this book. Surnames are more diverse than forenames, and there are some surnames for which I have been unable to find a ready Provençal equivalent; these I have left in Latin. Since Provençal was only infrequently a language of record, standardized vernacular spellings had not emerged. The area known as Cavalhon, for example, was also commonly called Cavallione, and the woman's name Beatrix in Latin had several different forms in Provençal; I have privileged one name, Cavalhon and Biatris in these instances, but other choices were possible. In general, I have used the most common spelling, although my choices are based on impressions, not exact word counts. Some elements of medieval Marseille's cartography have well-known modern French equivalents, and to avoid confusion I have left these in French. One example is the church of Notre Dame des Accoules whose spire survives to this day; in Provençal, the church was known as *las Accolas.* In toponyms containing place nouns such as "street" or "burg," I have capitalized the proper noun but not the place noun, as in "the street of the Corders." To capitalize place nouns is to give them an official or ontological status that they had not yet acquired in fourteenth-century Marseille.

A particularly thorny problem has been certain Latin or Provençal expressions that denoted neighborhoods or vicinities. The term *Fabraria* could be translated in English as "The Quarter of the Smiths," but since the original term does not contain the word "Quarter" the translation does not convey the sense of the original. In particular, "smith" itself is reduced to the object of a prepositional phrase and thereby seems to lose some of its ontological status in translation. The best translation of *Fabraria,* clearly, is "Smithery." Some of the vicinities in Marseille can be translated relatively smoothly in this way, such as "Tannery" and "Cobblery." "Smithery" at least makes sense in English, and others do not come off too badly, such as "Hattery" or "Nettery." Other translations, however, are less felicitous, such as "Jewelery," "Carpentery," and, worst of all, "Shoemakery." The fact that English and some other modern languages have, for the most part, lost this type of topographical expression nicely illustrates an argument made in the following pages.

In the later middle ages, the new year normally began on March 25, not on January 1. I have changed all dates to the new style.

Abbreviations

ADBR   Archives Départementales des Bouches-du-Rhône
AM   Archives Municipales de la Ville de Marseille
Inventaire Département des Bouches-du-Rhône. Ville de Marseille.
            Inventaire sommaire des archives communales antérieures à
            1790. Série BB. Vol. 1. Marseille, 1909.
Statuts   Régine Pernoud, ed. *Les statuts municipaux de Marseille.*
            Monaco: Imprimerie de Monaco, 1949.

IMAGINARY

CARTOGRAPHIES

Imaginary Cartographies

The margins and covers of the codices and cartularies of medieval Europe yield a rich harvest of images, since notaries and scribes were wont to doodle. If you poke around in the rich array of sources from fourteenth-century Marseille, you will find a study of profiles, noses, and keys. You will find saints and monsters. Next to a dotal act from 1348, celebrating life in that year of plague, you will see the veil of a virgin, and, peering out, her coal-black eyes. The inside cover of one casebook from the period is richly illustrated with a beautiful set of compass drawings. There is more, and the images are a pleasant distraction for the reader of the words. But never, in tens of thousands of pages of documentation, in countless acts providing addresses and boundaries, in immense rent registers that collectively list the locations of thousands of pieces of property, will you ever find an image even remotely like a map.

It is a remarkable documentary silence, and one that is not particular to Marseille. Medieval western Europe was one of the most cartographically inclined of all premodern worlds; only Arab geographers produced a corpus of maps that may have been similar in bulk to that of medieval Europe.[1] Yet the near-universal interest in producing *mappaemundi* in medieval Latin Christendom contrasts sharply with the local maps of the Islamic and Chinese corpus.[2] With some notable exceptions—such as a

[1] Islamic maps are discussed in the contributions found in J. B. Harley and David Woodward, eds., *Cartography in the Traditional Islamic and South Asian Societies*, vol. 2, bk. 1 of *The History of Cartography* (Chicago, 1987).
[2] On regional maps in China, see Mei-Ling Hsu, "An Inquiry into Early Chinese Atlases

few representations of several cities, various pilgrimage maps, the maps
Matthew Paris produced between 1236 and 1251, and portolan charts
beginning in 1290 with the *Carta Pisana*—there were very few attempts
to produce local maps of any kind in Europe before 1300. The field of
vision rarely shrank even to national or regional levels, in part because
the tangled array of feudo-vassallic rights and obligations made a unified
polity with fixed boundaries difficult to conceive. In the realm of law,
lordship, and administration, where the usefulness of representing properties, boundaries, and personal addresses by means of a graphical map
seems obvious to us, there is documentary silence.[3]

Most relevant to the matter at hand, urban maps—that is to say, urban
representations from the perpendicular that take notice of streets—
although common enough in the Roman empire, are almost entirely non-
existent in any part of western Europe before the fifteenth century.[4]
There were, of course, innumerable medieval images of cities, and there
is no clear boundary between art form and map or plan. In medieval
images, cities are often represented schematically by their walls and towers
or in profile views with buildings shown in elevation. Obscured by these
buildings are the streets. A favored object of early modern chorographies
and ichnographic plans, streets are rarely seen before 1500.[5]

---

through the Ming Dynasty," in *Images of the World: The Atlas through History*, ed. John A. Wolter
and Ronald E. Grim (Washington, D.C., 1997).

[3] The existing maps are surveyed by P. D. A. Harvey, *The History of Topographical Maps:
Symbols, Pictures and Surveys* (London, 1980); see also idem, "Local and Regional Cartography in Medieval Europe," in *Cartography in Prehistoric, Ancient, and Medieval Europe and
the Mediterranean*, vol. 1 of *The History of Cartography*, ed. Harley and Woodward (Chicago,
1987), 464–501. On the lack of maps and the unmappability of medieval power, see Robert
Fawtier, "Comment le roi de France, au début du XIVᵉ siècle, pouvait-il se représenter son
royaume?" in *Mélanges offerts à M. Paul-E. Martin par ses amis, ses collègues, ses élèves*, vol. 40 of
Mémoires et documents publiés par la société d'histoire et d'archéologie de Genève (Geneva, 1961),
65–77; Gérard Sivéry, "La description du royaume de France par les conseillers de Philippe
Auguste et par leurs successeurs," *Le moyen âge* 90 (1984): 65–85; Elizabeth Hallam, *Capetian France, 987–1328* (London, 1980), 78–86; Roger J. P. Kain and Elizabeth Baigent, *The
Cadastral Map in the Service of the State* (Chicago, 1992), 205–10; Gabrielle M. Spiegel, *Romancing the Past: The Rise of Vernacular Prose Historiography in Thirteenth-Century France* (Berkeley,
1993), 19. See also Bernard Guenée, "La géographie administrative de la France à la fin du
moyen âge: élections et bailliages," *Le moyen âge* 67 (1961): 293–323.
[4] Naomi Miller discusses some of these issues in "Mapping the City: Ptolemy's *Geography* in
the Renaissance," in *Envisioning the City: Six Studies in Urban Cartography*, ed. David Buisseret
(Chicago, 1998), 34–74.
[5] See Pierre Lavedan, *Représentations des villes dans l'art du moyen âge* (Paris, 1954), especially
his comments on "vues verticales" and streets, 36–38; John A. Pinto, "Origins and Development of the Ichnographic City Plan," *Society of Architectural Historians Journal* 35 (1976):
35–50; Juergen Schulz, "Jacop de' Barbari's View of Venice: Map Making, City Views, and
Moralized Geography before the Year 1500," *Art Bulletin* 60 (1978): 425–74; Harvey, "Local

This silence is rather surprising, since, as many have observed, maps are not only conveniences, but also instruments and reflections of power.[6] If maps are efficient instruments of the power of kings, states, or lords, they also carry a different sort of power, because they classify and select according to principles that may have little to do with politics and more to do with such things as the claims of Christendom, the values of civil society, or the principles of a scientific paradigm. To acknowledge the rich relationship between mapping and power is to realize the extraordinary significance of the *mappaemundi*, for they betray the very early origins of the European claim to global hegemony. But here is a conundrum: in medieval Europe we have audacious world maps, inherently didactic and Christian, but little that was regional or local, royal or seigneurial, little to indicate that the vast body of people who lived outside the intellectual centers of the Christian church were at all aware of the many ways in which maps configure the world. In the thousands of pages of documents from medieval Marseille that concern property rights, you will never find a graphic representation of space, and this despite the understanding, common since the twelfth century, that pictures were economical ways to signify things written.[7] Even *terriers*, great medieval registers or rolls enumerating rights in land, inherently topographical in nature, include maps only rarely.[8] In the archives of Marseille, one has to travel down the ages

---

and Regional Cartography," 476–82; Chiara Frugoni, *A Distant City: Images of Urban Experience in the Medieval World*, trans. William McCuaig (Princeton, 1991); Thomas Frangenberg, "Chorographies of Florence: The Use of City Views and City Plans in the Sixteenth Century," *Imago Mundi* 46 (1994): 41–64.

[6] An argument closely associated with the work of the historian of cartography J. B. Harley; see, among his many works, "Maps, Knowledge and Power," in *The Iconography of Landscape: Essays on the Symbolic Representation, Design, and Use of Past Environments*, ed. Denis Cosgrove and Stephen Daniels (Cambridge, 1988), 277–312. See also Michel Foucault, *Power/Knowledge: Selected Interviews and Other Writings, 1972–77*, ed. Colin Gordon, trans. Colin Gordon et al. (New York, 1980), 63–77; "Maps of Authority: Conflict in the Medieval and Early Modern Urban Landscape," special number of the *Journal of Medieval and Early Modern Studies* 26 (1996); Jeremy Black, *Maps and Politics* (London, 1997).

[7] M. T. Clanchy, *From Memory to Written Record: England, 1066–1307*, 2d ed. (London, 1993), 288–91. R. A. Skelton suggests that Albertus Magnus and Roger Bacon were largely responsible for "the recognition that a graphic design will communicate geographical relationships more efficiently than a written document, and that the elements of a map can only be transmitted by text or by word of mouth with great difficulty. See Skelton, *Maps: A Historical Survey of Their Study and Collecting* (Chicago, 1972), 7.

[8] Harvey, "Local and Regional Cartography," 464–65, 470; Kain and Baigent, *Cadastral Map*, 3–6. Notaries elsewhere in Europe could and occasionally did produce maps in an effort to clarify property locations; see *Il notaio nella civiltà fiorentina: secoli XIII–XVI: mostra nella biblioteca medicea laurenziana, Firenze, 1 Ottobre–10 Novembre 1984* (Florence, 1984), 300. To the best of my knowledge such maps are rare.

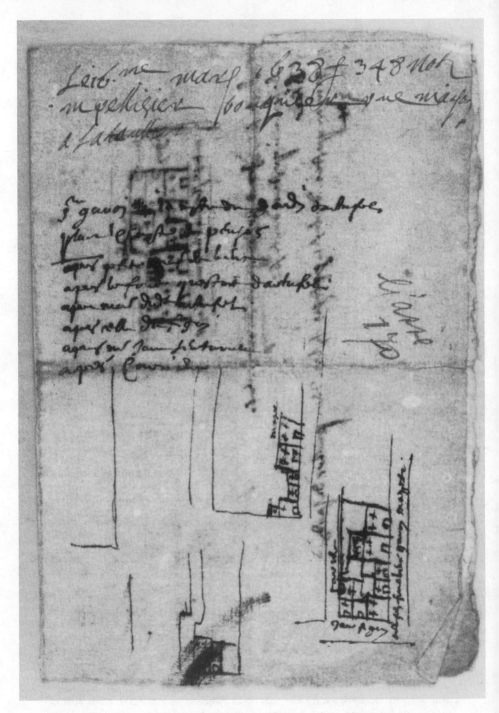

1. Earliest mapping of episcopal lordship in Marseille, ca. 1638.

to the seventeenth century to find the first graphic representation of the space of lordship—scraps of paper, crudely done, now floating in a mass of irrelevant documents related to episcopal rents (Figure 1).[9]

By this time, of course, western Europe had already experienced phenomenal growth in cartography. By 1600 maps were everywhere in use, representing a diverse array of subjects—nations, trade routes, linguistic regions—and responding to demands ranging from military concerns to tourism.[10] An increasingly sophisticated mathematics of cartography generated maps from a multitude of perspectives, most notably local and regional maps and, eventually, cadastral maps. New technologies, like the printing press and copper-plate engraving, contributed to the rapid increase in map production.[11]

At one time, the rapid growth in cartography and the increasing accuracy of maps was explained as one of the most perfect indices of the scientific revolution.[12] More recent interpretations of the early modern *furor geographicus* go beyond the tools of production and search for answers rooted in state-building and identity formation.[13] A recent collection opens with the query: "[H]ow did it come about that whereas in 1400 few people in Europe used maps, except for the Mediterranean navigators with their portolan charts, by 1600 maps were essential to a wide variety of professions?" The succinct answer is captured in the title of the collection: "Monarchs, Ministers and Maps: The Emergence of Cartography as a Tool of Government in Early Modern Europe."[14] Here, it is the hegemonic discourse within which a given map is framed that becomes

---

[9] ADBR 5G 170, liasse 170, ca. 1638. See also 5G 156 from later in the century, in which numerous maps can be found.

[10] On tourism, see Frangenberg, "Chorographies of Florence," 47–52; on military concerns see idem, 45–47, and P. D. A. Harvey, *Maps in Tudor England* (Chicago, 1993), 26–41.

[11] David Woodward, ed., *Five Centuries of Map Printing* (Chicago, 1975).

[12] On this see David Turnbull, "Cartography and Science in Early Modern Europe: Mapping the Construction of Knowledge Spaces," *Imago Mundi* 48 (1996): 5–24.

[13] In addition to some of the citations in note 6, see Harvey, *Maps in Tudor England*, esp. chap. 3, "Maps and Government," 42–65; Kain and Baigent, *Cadastral Map*; Josef Konvitz, *Cartography in France, 1660–1848: Science, Engineering, and Statecraft* (Chicago, 1993); David Turnbull, "Constructing Knowledge Spaces and Locating Sites of Resistance in the Modern Cartographic Transformation," in Rolland G. Paulston, ed., *Social Cartography: Mapping Ways of Seeing Social and Educational Change* (New York, 1996), 53–79. Recent work, like Peter Sahlins's argument that the idea of a boundary was generated not at the center of society and government but rather at the boundary itself, has questioned elements of this relation between power and cartography, but Sahlins, like Thongchai Winichakul, nonetheless suggests that mapping and boundaries are intrinsic to the political conditions of modernity. See Peter Sahlins, *Boundaries: The Making of France and Spain in the Pyrenees* (Berkeley, 1989); Thongchai Winichakul, *Siam Mapped: A History of the Geo-Body of a Nation* (Honolulu, 1994).

[14] David Buisseret, ed., *Monarchs, Ministers and Maps: The Emergence of Cartography as a Tool of Government in Early Modern Europe* (Chicago, 1992).

the object of study, and what is not represented by cartography becomes as significant as what is.[15]

Whatever the interpretation, the profusion of early modern maps serves to underscore the apparent medieval failure to map anything smaller than the entire globe. It is hard to resist the conclusion that the cartographic silences of medieval documentation reveal a primitive political and scientific culture, one that cared little about political centralization or national identity, one that saw maps not as ways to represent space, power, and identity, but rather as ways to preserve the fantastic cosmology of the ancient world in suitably Christianized form. This, indeed, is more or less what some earlier historians of cartography do say. In 1966 R. A. Skelton observed that "the medieval world map was drawn from literary sources [i.e., Greco-Roman]; it was schematic in form and didactic in intention; and only very slowly, in the fourteenth and fifteenth centuries, was it to admit information from experience."[16]

Yet this was not necessarily some failure of imagination when measured against a modern standard of efficiency of representation. As evolutionary psychologists and linguists point out, space itself is a fundamental metaphor in all human languages, and is therefore deeply ingrained in the mental apprehension of the world.[17] The brain works spatially, whether it complements words with graphic images or relies on words and ideas alone. In the more immediate realm of social experience, every child learns how to navigate space without using maps, and men and women, as speakers of a common language, naturally establish linguistic cartographic conventions. These conventions are plural, a point that is made clear whenever we have difficulty following another's verbal map, though they are conventions of mapping, nonetheless. More particular to this book, every one of the hundreds of thousands of property conveyances extant from medieval Europe included a clause identifying the location of the property by means of words alone. When, in 1352, a citizen of Marseille named Aycart Valhon purchased a house on the corner of two public streets, the notary who transcribed the transaction conscientiously named the two streets—"the avenue of the [gate of the] Frache

---

[15] See especially Richard Helgerson, "The Land Speaks: Cartography, Chorography, and Subversion in Renaissance England," *Representations* 16 (1986): 50–85; J. B. Harley, "Silences and Secrecy: the Hidden Agenda of Cartography in Early Modern Europe," *Imago Mundi* 40 (1988): 57–76.

[16] See Skelton, *Maps*, 6–7. Similar statements can be found in Lloyd A. Brown, *The Story of Maps* (Boston, 1949).

[17] Steven Pinker summarizes some of the literature on the subject in *How the Mind Works* (New York, 1997), 352–55.

and a street or set of steps which leads to the church of St. Martin"[18] —
thereby neatly locating his client's new possession, through a kind of tri-
angulation, at the intersection of both civic and ecclesiastical monuments.
This clue tells us something about the ways in which urban space in the
record-keeping bureaucracies of the later middle ages was, in fact,
mapped, although mapped linguistically rather than graphically. We
would need to study many more such clauses to figure out exactly how
this was done. Certainly this single description leaves much unsaid, and
its silences are meaningful. Where is the administrative district or the
parish in this plat description? Why doesn't the second street or alley bear
a name? Was this the notary's oversight, or his client's, or a general cartographic oversight? Was there a social or aesthetic hierarchy of public
ways, from avenue (*carreria recta*) to street (*carreria*) to alley (*transversia*)
to steps (*deyssenduda*)? From this single clue, without controlling documents, we cannot even tell whether this was the notary's cartography, the
client's, the property lord's, or some cartography universal to all residents
of Marseille. Only by looking at a number of these clauses can we begin
to develop a meaningful understanding of the mental cartography that
told interested parties where and how to locate Aycart Valhon's new
house.

Despite the total absence of graphic maps in medieval Marseille, we
can see that Aycart Valhon's house was mapped by means of an as yet
undefined imaginary cartography. Aycart himself was not "mapped"; that
is to say, he had no address. Originally from the nearby castrum of
Allauch, Aycart was identified in the act only as a citizen and resident of
Marseille. To be a citizen was a precise legal status that required both an
investment in the city and a period of residence of one year. But the
identification clause in this act does not give an address. Such clauses in
other acts do, as in the case of Isnart Draguet, who in March of 1350
acquired a house on the street of Cavalhon located "under the house of
Isnart Beroart"—another intriguing form of mapping space that differs
from that used in the case of Aycart Valhon's house—and in the act was
identified as a laborer living on the nearby street of Pons Broquier.[19] Use
of such an address, precise by the standards of a time that had some house
names but no house numbers, was fitful at best. Identity was not routinely
linked to residence in notarial records in medieval Marseille, and an
address, so central a part of identity in the classifying schemes of modern
record-keeping bureaucracies, was only just becoming thinkable by other

[18] ADBR 355E 5, fols. 27r–v, 11 June 1352.
[19] ADBR 355E 3, fols. 3v–4v, 29 Mar. 1350.

agents of record in the later middle ages. Yet like plat descriptions, the residential components of identity clauses, where found, also borrowed from an implicit linguistic cartography, and both clauses used this cartography to map out possession and identity.

Of course, to frame the subject in this way is to stretch the meaning of cartography. Maps, according to Harley and Woodward, are graphic representations.[20] Closer to the concerns of this book, historians of property mapping have taken the reasonable view that a cadastral map must be graphic and not linguistic in form, and for this reason they have not spent a great deal of time exploring the non-graphical maps that informed medieval legal documents. In an important survey entitled *The Cadastral Map in the Service of the State*, Roger J. P. Kain and Elizabeth Baigent suggest that the linguistic nature of what one might call medieval linguistic property maps makes them inefficient vehicles for state interest and power. For this reason, they do not include such linguistic maps in the genealogy of the early modern cadaster.[21] This is an important observation, but on some level it cannot be right. All maps, linguistic or graphic, are sites for the expression of power, state or otherwise, even if changing technologies of map production allow for changes in the expression of that power. However, Kain and Baigent are absolutely right on one score: what is particularly significant about late medieval linguistic property maps is that the power they express is not exclusively or even primarily state power. Agents of record-keeping bureaucracies in Marseille who worked for the state, namely, the Angevin kingdom in Naples, did create and use a linguistic cartography, especially for purposes of taxation. In Marseille, however, the cartography used by these functionaries was based on city blocks, an ancient cartographic form progressively being eliminated from late medieval cartographic thought. At least in Marseille, the linguistic roots of the cadastral maps studied by Kain and Baigent do not lie in a state-sponsored cartographic science. They lie elsewhere.

This book is about this cartographic science. I take as my subject the linguistic mapping of possession and identity in the later middle ages, and I consider property and people together because the technologies used to map the one are readily available for the mapping of the other. The basic sources used in this book are the site descriptions and personal identity clauses one finds in a wide variety of acts—notarial, seigneurial, fiscal— that identify where a unit of urban property or a person was located. Site descriptions and personal identity clauses are hardly unknown to medieval

[20] Harley and Woodward, eds., *Cartography in Prehistoric*, xix.
[21] Kain and Baigent, *Cadastral Map*, 3.

social historians, geographers, architectural historians, and archeologists, who have used them to reconstruct village and urban plans and to write histories based on information about where people lived, how they married, and how they imagined themselves within larger groups, such as kinship, confraternity, or social class.[22] Here, we shall be concerned less with the positivistic reconstruction of urban space in late medieval Marseille than with the mental maps that lay behind these descriptions.

Historians and archeologists who work with these sorts of documents have long known about, and been frustrated by, the non-standard nature of medieval identity clauses. Where personal identity clauses indicate public figures, such as kings or abbesses, more precise delimiting identifiers based on genealogy, legal domicile, and personal features do not seem entirely necessary from the standpoint of legal signification—everyone, after all, knows the person in question, in the same way that nowadays we know the person for whom we are voting even though delimiting identifiers are not used in the voting booth. On the level of private citizens and their property in the middle ages, however, it is a little surprising to realize that the language of identity clauses was not always precise and was often less than consistent. Clearly, the person or entity in question, despite a lack of public significance or notoriety, was still signified more by social memory than by verbal devices, in this case the memory of the agent of record. Since there was little pressure to improve on social memory, the language of record-keeping itself did not have, or need, the full range of categories used by bureaucracies today to frame identities.

To work around these basic facts is to avoid the interesting question of why there were so few categories of identity throughout much of the middle ages and why such categories gradually get more precise, and more abundant, over time. Since Aristotle, nominalists have argued that categories themselves do not reside in nature. Mary Douglas, summarizing the work of the philosopher W. V. Quine, noted that "identity or sameness is conferred on objects by their being held in the embrace of a

---

[22] The field of historical geography is too vast to survey here; see, among others, Dietrich Denecke and Gareth Shaw, eds., *Urban Historical Geography: Recent Progress in Britain and Germany* (Cambridge, 1988). Useful representatives of the work of historians include Samuel Kline Cohn, *The Laboring Classes in Renaissance Florence* (New York, 1980); Louis Stouff, "Arles à la fin du moyen âge. Paysage urbain et géographie sociale," in *Le paysage urbain au moyen âge* (Lyon, 1981), 225–51; idem, "La population d'Arles au XVᵉ siècle: composition socio-professionnelle, immigration, repartition topographique," in Maurice Garden and Yves Lequin, eds., *Habiter la ville, XVᵉ–XXᵉ siècles* (Lyon, 1985), 7–24; Derek Keene, *Survey of Medieval Winchester*, 2 vols. (Oxford, 1985); Jean-Claude Maire Vigueur, ed., *D'une ville à l'autre: structures matérielles et organisation de l'espace dans les villes européennes (XIIIᵉ–XVIᵉ siècle)* (Rome, 1989).

theoretical structure," or, as Douglas would argue, in the embrace of institutions, one of whose roles is to confer sameness.[23] This observation encourages us to study the historical development of the institutions that were in the process of refining the categories of identity used in records.[24]

In medieval Marseille, we find considerable uniformity on the level of names. Almost all men, women, and children in Marseille from the thirteenth century onward bore a given name and a surname. This is not typical of all naming patterns in Europe, as in many parts of Italy and England, where surnames are relatively uncommon in the thirteenth century.[25] Beyond the relatively uniform naming pattern in Marseille, however, lies an array of additional identifying labels that includes all possible permutations of kinship, status, and domicile, including no label at all. Clearly, no template established by late medieval chancery or even local notarial practice existed to guide identifications, and the result was an erratic selection shaped by particular circumstances or, in some cases, by habits particular to status groups. Individual expressions of identity pushed through the record-keeping format far more then than they do now. Above all, within the context of place of residence or plat location, we find fundamental disagreements on what should be the cartographic norm. In the same way that there existed no personal identity template to systematically guide the process of shaping a personal identity, so too there existed no universal cartographic template according to which addresses and plats were systematically located in space.

There are, of course, numerous ways of mapping the world, and perhaps this inconsistency should not surprise us. All maps have a lexicon or a set of toponyms, and these toponyms will vary according to the goals of the cartographer. Toponyms, in turn, are organized on a map according to some reference grid or cartographic grammar. Graphical maps are

[23] Mary Douglas, *How Institutions Think* (Syracuse, 1986), 59. Douglas summarizes arguments made by W. V. Quine in *From a Logical Point of View* (New York, 1961). Some cognitive linguists and cognitive scientists agree that categories do not reside in nature, although, unlike Quine and Douglas, their explanations center on cognition and not on institutions; see George Lakoff, *Women, Fire, and Dangerous Things: What Categories Reveal about the Mind* (Chicago, 1987); see also the more muted position summarized by Steven Pinker in *How the Mind Works*, 126–29, 306–13.

[24] A similar interest in how social categories historically acquire definition emerges in R. I. Moore, *The Formation of a Persecuting Society: Power and Deviance in Western Europe, 950–1250* (New York, 1987), and Michel Foucault, *Madness and Civilization: A History of Insanity in the Age of Reason*, trans. Richard Howard (New York, 1965).

[25] Consider, for example, the city of Florence. See David Herlihy and Christiane Klapisch-Zuber, *Les toscans et leurs familles: une étude du catasto florentin de 1427* (Paris, 1978), 539.

fixed in two dimensions according to rules of perspective and scale and a chosen set of reference points. These unite to produce a multitude of possible maps. We are not surprised when a political map of a given region looks entirely different from a map of its vegetation or language groups, because we use these maps for different purposes. There is a linkage between the interests of the cartographer and the grid chosen to convey that interest.

Indeed, it is this linkage that explains the inconsistencies in the linguistic map of medieval Marseille. The variations are not entirely random, for particular linguistic communities tended to favor particular cartographic grids. As I will use the term here, "linguistic community" refers to a group of people who converse and share and remember information together on a regular basis, and therefore by virtue of this conversation and shared knowledge develop linguistic conventions and categories that are particular to the group in question. I use "linguistic community" in preference to the term "speech community," commonly found in sociolinguistic literature, because the latter is inappropriate in a context in which linguistic conventions that can be studied are written and not spoken.[26] Membership in one linguistic community, obviously, does not preclude membership in another, and by using the expression I do not wish to reify the groups in question *a priori*. Three such communities interest me most. Two, namely public notaries and seigneurial officials, kept records in Latin. To a certain extent the community of these two groups must be thought of as including all prior agents of record whose linguistic conventions shaped the language of the documents used by all succeeding generations. A third group consists primarily of non-noble speakers of Provençal, the vernacular language of medieval Marseille. These linguistic communities are in no way commensurate in size and do not correspond neatly to social groups; I identify them solely by their tendency to share certain cartographic conventions. The communities so defined might look very different if the principle chosen were something other than cartography. Yet I do think, and seek to show, that a shared cartography reflects shared interests and certain intangible political goals.

---

[26] See J. J. Gumperz, "The Speech Community," in *Language and Social Context*, ed. Pier Paolo Giglioli (Harmondsworth, 1971), 219–31; see also Suzanne Romaine, *Socio-Historical Linguistics: Its Status and Methodology* (Cambridge, 1982), 234–38; Glyn Williams, *Sociolinguistics: A Sociological Critique* (London, 1992), 69–75. For general remarks regarding the historical application of sociolinguistics, see Romaine, *Socio-Historical Linguistics*, 15–21; Peter Burke and Roy Porter, eds., *The Social History of Language* (Cambridge, 1987), 1–20; Peter Burke, *The Art of Conversation* (Ithaca, 1993), 1–33.

The verbal maps of property sites and addresses generated by these three linguistic communities generated four distinct cartographic templates and several others that were, statistically speaking, less important; and, although each linguistic community borrowed freely from all available templates, each nonetheless showed distinct preferences for one or at most two. The first and most important template was based on streets or related entities like alleys and plazas; this was the template that would, in time, become the urban norm. Here, the street is the chief point of reference; both addresses and property sites are given as streets, and the map of the city, accordingly, is reduced to a simple architectonic form. Although all people in Marseille understood and talked about streets, the template was conspicuously favored by the public notaries of the city. Another template took city blocks or, as the records call them, "islands" (*insulae*) as the basic cartographic point of reference. According to this template, streets fade in significance; they serve only to carve out islands of houses and gardens. One surprising consequence is that streets, in the insular context, often lose their names. Moreover, the insular template was used almost exclusively in records kept by two powerful direct lords, the bishop and the crown. The city council also used the language of islands in certain fiscal records.

Both the street and the insular templates were frequently used by agents of record in mid-fourteenth-century Marseille. Both were Latin and therefore learned cartographic templates. Neither, in the fourteenth century, was in any way official, nor is there any particular reason to think that Latin cartographies of record were inherently simple translations of vernacular linguistic cartographies. There can be a considerable difference between the way a city is mapped by its bureaucracy and the way it is mapped by its residents. In a contemporary context, consider Roland Barthes' observations of Tokyo.

> The streets of this city have no names. There is of course a written address, but it has only a postal value, it refers to a plan (by districts and by blocks, in no way geometric), knowledge of which is accessible to the postman, not to the visitor: the largest city in the world is practically unclassified, the spaces which compose it in detail are unnamed. This domiciliary obliteration seems inconvenient to those (like us) who have been used to asserting that the most practical is always the most rational. . . . Tokyo meanwhile reminds us that the rational is merely one system among others.[27]

---

[27] Roland Barthes, *Empire of Signs*, trans. Richard Howard (New York, 1982), 33.

In contrast to modern day Tokyo, it is not easy to know how to go about recovering the vernacular linguistic cartography of medieval Marseille, if indeed there was such a thing, and if it differed from the cartographies of agents of record. Almost by definition, the records from the later middle ages were those kept in Latin by record-keeping bureaucracies. All the same, a few records in Provençal have survived from late medieval Marseille, and notaries and seigneurial agents themselves occasionally use templates that are neither those of street or island. From these sources we can construct the outlines of the vernacular linguistic cartography, that is to say, the cartography most often used by ordinary speakers of Provençal.

If vernacular cartography in late medieval Marseille borrowed freely from all available templates, it was characteristically based on the templates of vicinity and landmark. To begin with the former, a vicinity, unlike a street or an island, was not a named concept. The word "vicinity" (Lat. *vicinia*) does appear from time to time in contemporary records, but whereas people will speak of property located in the "street of the Oarmakers" (*carreria Remeriorum*) or the "island of the Oarmakers" (*insula Remeriorum*), no one ever spoke of a hypothetical "vicinity of the Oarmakers" (*vicinia Remeriorum*). This does not necessarily imply that vicinities had lower ontological status than streets or islands, and as I will argue later, vicinities are clearly marked in grammar if not in lexical usage.[28] It is possible that people did not adapt the word *vicinia* and its Provençal cognates for use in a template precisely because a vicinity is not a readily mappable concept. We cannot represent it graphically without subverting the purpose it served. This is because the vicinity was a space of sociability. When artisans, merchants, retailers, members of service trades, professionals, laborers, fishermen, and even notaries contemplated the city and talked about its map in ordinary Provençal they often "saw" not a skeletal tracery of streets, much less a collection of islands, but floating knots of residential sociability or centers of production or retailing that extended across several streets, alleys, or intersections. The city dwellers gave these vicinities as their addresses and, then as now, such addresses carried status and significance.

In addition to vicinities, ordinary speakers of Provençal also saw landmarks—notable men and women, churches, ovens, gates, fountains, baths, features of the city's physical topography, inn or house signs, ghet-

---

[28] As William Ian Miller notes, the absence of a word for "feud" in medieval Icelandic cannot possibly mean that medieval Icelanders had no concept of feud. See his *Bloodtaking and Peacemaking: Feud, Law, and Society in Saga Iceland* (Chicago, 1990), 181–82.

toes, such as the Jewry and the Prostitutes' quarter, images or statues—
and either confected these into vicinities or left them as landmarks. The
difference often lies in the preposition of choice. Typically, one lived at
or in a vicinity. In contrast, one lived next to, across from, close by, under,
or above landmarks. Used with relational prepositions like these, landmarks constitute a fourth template, although the template was very close
in usage to the vicinal template. It is worth observing here that in the
fourteenth and fifteenth centuries, at least in Marseille, one never lived
north, south, east or west of landmarks, and the custom of saying that one
lived above or below a landmark was in no way a reflection of the north-
south orientation of the Ptolemaic map and its modern descendants. If
one lived "below the hospital of St. Esprit," that was because St. Esprit was
perched on a slope that descended gradually toward the harbor. One
lived, quite literally, below it.

As this suggests, in vernacular linguistic cartography the city was made
of people and the landmarks that impinged on their consciousness. This
being so, we can appreciate why the men, women, and children whose
daily itineraries animated the streets of medieval Marseille used both
people and landmarks in their navigational conventions.[29] In a few
legal documents defining boundaries that have survived from medieval
Marseille, the boundaries unroll as if seen by a videocamera; one literally
walks through the city with the narrator as he takes you in 1357 from the
house of Carle de Montoliu near the Jewry to the house of a Jewish dyer,
past the drains of the old Jewish wells, thence to the gate where the Jew,
Durant de Bedarides, lives, and finally to the alley that leads to the church
of St. Martin where Bondavin's garden is.[30] One navigates as if through a
crowd of people and landmarks, not along a tracery of streets. Although
the architectonic form of streets existed in medieval Marseille, people just
did not "see" it in much the same way that medieval artists, when por-
traying a city, saw buildings in elevation in place of streets. This has an
interesting corollary: since people did not see them as such, streets in
France were often unnamed before the thirteenth century. As Jean-Pierre
Leguay remarks,

---

[29] Travel literature and similar sources give us some idea of how people in the middle ages
navigated their way through space, and the space through which they navigated was very
different from the space that can be imagined today by anyone who thinks with maps. See
esp. Jean Boutier, Alain Dewerpe, and Daniel Nordman, *Un tour de France royal: le voyage de
Charles IX (1564–1566)* (Paris, 1984).

[30] See AM BB 22, fol. 27r. The text refers to a boundary set up in the Jewry to demarcate
a living space for Christians forced to flee the suburbs because of the threat of an invasion.
See also the text of an accord between the rectors of the commune of Marseille and
the bishop of Marseille from 23 January 1220, ADBR B 315, transcribed in Bourrilly, *Essai*,
315–25.

The appearance of a name is a concrete expression of the importance accorded to a street: the density of its population, the significance of its role as an access, or its economic function. [It is evidence for a] burgeoning urbanization if it concerns an existing rural path recently absorbed by the expansion of the city. Unhappily, the attempt to employ a study of names for the purpose of identifying streets will not yield results until the thirteenth century and beyond. Before, the alternative is the rule.[31]

Vernacular linguistic cartography, then, was dependent on social context. Both the templates and the toponyms used in vernacular linguistic cartography identified living features of the city's physical, artisanal, commercial, and social landscape. As a result, toponyms and grammars changed as this landscape evolved, as people died and buildings were torn down and the memories of them were effaced, as landmarks evolved into vicinities. Linguistic conventions existed in a given generation, but the absence of an official map ensured that the map itself was mutable.

There were no rules or customs determining cartographic usage in any linguistic community in late medieval Marseille. All four major templates were available for use by everyone. Although vernacular linguistic cartography paid no attention to islands, it did take heed of streets and at times borrowed some of the districts that figured in administrative cartography. Notaries favored streets but the whole gamut of possible templates surface in the property site clauses they drew up. Episcopal, royal, and municipal officials preferred islands but borrowed from other templates from time to time. Sometimes such borrowing was necessary. Not all parts of the city, for example, could be readily mapped by means of islands, such as isolated houses that fell under the seigneurie of a certain lord, and therefore seigneurial officials occasionally borrowed from the templates of street or landmark to describe these ill-defined spaces. It is important to bear in mind that the preferences of a given linguistic community were a matter of percentages, not absolute norms.

The templates themselves, however, were mutually exclusive. They conveyed different ideas about what should be the basic unit of the address or plat description. Hence, one does not often find a plat description or an address in medieval Marseille defined according to two or more templates, as in a hypothetical "located in the Tilery in the street of the Tilery." The major exceptions are plat descriptions in which one cartographic term nests inside another, as in this legitimate phrase: "located in the street of the Tilery in the burg of St. Augustine."

---

[31] Jean-Pierre Leguay, *La rue au moyen âge* (Rennes, 1984), 92.

Cartographic templates in medieval Marseille usually shared the same lexicon of toponyms. In the mid-fourteenth century, for example, notaries typically referred to a certain place near the port as the street of the Changers, the *carreria Cambiorum*. Provençal speakers preferred to call it the Change, or the *Cambio*, as it was known in Provençal. Scribes associated with powerful landlords invariably called it the island of the Changers, the *insula Cambiorum*. Here the lexical term remains the same. It is the structure of the space—imagined as a street, vicinity, landmark, or block—that shifts. Although this appears to be a rather trivial linguistic distinction, it is significant for several reasons. First, members of these three groups were, in the appropriate context, fairly consistent in following the style of their group. This suggests either a degree of collective consciousness or the presence, within the conversational circles of these groups, of governing conventions or linguistic norms on cartographic matters. Second, there is a correlation between the social or political goals of the three groups and the different templates each tended to prefer. This is especially clear in the case of artisanal groups or retailers, whose language suggests that they saw their vicinities as autonomous and self-reliant entities.[32] This is not a trivial point. As John J. Gumperz and Dell Hymes have argued, "all speech communities are linguistically diverse and it can be shown that this diversity serves important communicative functions in signaling interspeaker attitudes and in providing information about speakers' social identities."[33] According to this argument, vernacular cartographic categories reflected vernacular social identities. If this is true, then one can appreciate why the notaries preferred a template whose basic grammar dissolved these vicinities: this was an indirect assertion of power, a strategy that encouraged people to rely on the mediating role of notaries. A similar situation may also hold with the insular template, although it is less clear how the insular template might have fostered the political goals of seigneurial agents.

The four major templates discussed above do not exhaust the possible cartographic grids of the city. After all, linguistic communities mapped out more than just possession and identity, and cartographic grids used for other purposes were different. Notable among these was the political map of the city. In certain fiscal records identifying tax obligations, we

[32] James S. Amelang notes a similar situation in early modern Barcelona; see his "People of the Ribera: Popular Politics and Neighborhood Identity in Early Modern Barcelona," in *Culture and Identity in Early Modern Europe (1500–1800)*, ed. Barbara B. Diefendorf and Carla Hesse (Ann Arbor, 1993), 119–37.

[33] John J. Gumperz and Dell Hymes, eds., *Directions in Sociolinguistics: The Ethnography of Communication* (New York, 1972), 13.

find that royal and municipal administrators used a fairly elaborate political map of the city that was based on administrative divisions. Inside the walls of the city we find administrative quarters or *sixains*.[34] Outside the walls of the city, we hear of suburbs or burgs. The church, in turn, divided the city according to parishes. Elsewhere in medieval Europe—especially, it seems, in places where political power was more thoroughly centralized than it was in Marseille—both parishes and administrative quarters were incorporated into site descriptions and often constituted the lowest level of address used by citizens and classifying agents alike. In the case of fiscal and judicial records of Italian cities, Julius Kirshner has observed that these "almost always included a person's town, neighborhood, or at least a toponymic surname."[35] Part of the impetus for including the town came from the importance of citizenship in the *ius commune*; neighborhood or section, in turn, was important in fiscal contexts. According to Derek Keene and Vanessa Harding, properties in London between the twelfth and late sixteenth century were usually defined by reference to parishes, although sometimes streets or house names were used instead.[36] The street begins to systematically replace the parish as the form of address in London by the seventeenth century.[37] These are not notarial records, of course, and we cannot assume that strategies for mapping identities hold across all types of records—fiscal, judicial, seigneurial, notarial, and so on. In Marseille, for reasons that cannot easily be explained, notaries, seigneurial and municipal officials, as well as ordinary men and women, were exceedingly resistant to using administrative quarters and parishes in their addresses. Ordinary residents of the city spoke of the burgs and certain informal districts, but their selective patterns of usage suggest that they saw these as larger versions of vicinities. So rare is the word "parish" (*parochium*) in extant documents, including even those documents emanating from the episcopal curia or the cathedral canons, that we are completely unable to reconstruct the boundaries of the parishes of fourteenth-century Marseille. Although systematic comparative evidence is lacking, the tendency to avoid districts appears unique to Marseille, in sharp contradistinction to many cities of the Italian peninsula, where quarter identity was deep. As Noël Coulet has argued, fourteenth-century Marseille was possibly distinctive even among Provençal cities and

---

[34] Notable among these is the general taille of 1360–1361; see AM EE 55A. Officials in the individual *sixains* were responsible for maintaining a list of their assessable residents, and hence taxpayers in this register were identified according to their *sixain* of residence.

[35] Personal communication, Sept. 1998.

[36] Derek Keene and Vanessa Harding, *A Survey of Documentary Sources for Property Holding in London before the Great Fire* (London, 1985), xv.

[37] Derek Keene, personal communication, Sept. 1998.

towns.[38] This is but one suggestion that cartographic templates were sensitive to local political conditions. Apparently, templates favored by centralizing institutions were unfavored in the relatively decentralized political conditions of late medieval Marseille.

At a certain level of abstraction, even street, vicinity, landmark, parish, quarter, and burg do not exhaust the possible cartographic grids. Among other things, there are good reasons to believe that people who styled themselves as "nobles"—a group of notable families and individuals defined loosely by the tendency of its members to use distinctive titles and to own rents and rural estates, by their distinctive patrimonial inheritance practices, by an awareness of their genealogy, and by their reluctance to pay direct taxes—mapped the city in lexically distinctive ways, even if their templates followed the norms of other groups. Unfortunately, there are few sources from which we might begin to systematically measure the distinctiveness of the nobles' lexicon of toponyms. Various clues found in the testimony of noble witnesses in court cases do suggest that the cartographic conversations of the nobility may have developed a set of toponyms, derived in large part from the names of noble families, that were not shared by other city residents. Where nobles spoke of the "plaza of the Vivaut," for example, others spoke of a vicinity called "Pilings" that took its name from pilings set in the quay constituting one side of the plaza.

Perhaps most interesting about the cartography of the nobility is the tendency for members of noble factions to map the city according to lines of hatred, an emotional mapping that complements the moral cartography implicit in the vicinity. When members of the two noble factions testified in court about the emotional context of their violent exchanges, they painted an image of the city and its neighborhoods as polarized by hatred. In their testimony, the defendants in the trial that followed on the heels of the great battle of 1351 created a mental geography that carefully partitioned the city into territories controlled by the Vivaut and the Jerusalem and by their allies. From testimony given at the inquest, it is clear that the invading Jerusalem were thought to have offended the residential space of the Vivaut. As a Vivaut adherent named Bertomieu Bonvin was participating in a council to discuss what to make of the peace offered to them, he heard the cry "A las armas! a las armas!"[39] Hastily arming himself, he rushed to the corner of the street of Guilhem Tomas which he described as "inside their boundaries" (*intra eorum confinias*). He

---

[38] Noël Coulet, "Quartiers et communauté urbaine en Provence (XIIIᵉ–XVᵉ siècles)," in *Villes, bonnes villes, cités et capitales. Études d'histoire urbaine (XIIᵉ–XVIIIᵉ siècle) offertes à Bernard Chevalier*, ed. Monique Bourin (Tours, 1989), 357.

[39] ADBR 3B 811, fol. 30v.

was not alone in speaking of boundaries.[40] Depositions from other trials also make clear that territory was considered a convenient way to identify people. Asked how he was able to identify two men as members of the Vivaut party, a Jerusalem associate testified that he "saw them daily in the plaza of the Vivaut with arms."[41]

As members of the factions walked through the city, therefore, they saw neighborhoods polarized by hatred. Such was the case with Esteve de Brandis, later in life one of Marseille's greatest merchants, but in his youth a vigorous supporter of the Vivaut faction. Here's how the notary of the court of inquest transcribed his testimony concerning his actions on the day of the great battle of 1351:

> He was walking through the Cobblery on the day of the battle. When he came before Antoni Guigo's house, Lois de Tos, fully armed with every sort of weapon, chased him up to the stone where the water laps. Seeing that he couldn't remain safely in his house, he left to go to the plaza of the Vivaut for his own safety, and especially to that place because Johan Vivaut is his affine.[42]

Antoni Guigo was a Jerusalem adherent. The Cobblery, located a block or two east of the street of the Jerusalem, was Jerusalem territory. The plaza of the Vivaut further to the west was much safer ground.

What is perhaps most interesting about the geography of hatred was the way it rooted itself in Marseille's toponymy. The plaza of the Vivaut and the street of the Jerusalem endured for centuries, and although the street of the Jerusalem eventually faded (in the sixteenth or seventeenth century), you can still see, in modern Marseille, a street sign bearing the name "Place des Vivaut." Hence to conclude, although the four major templates are by far the best documented and therefore constitute the bulk of the source material used in this book, they do not, for all that, exhaust the possible ways of framing space in late medieval Marseille.

All the same, the world of diverse cartographies as it existed in late medieval Marseille was not a static world. Over the course of the fourteenth, fifteenth, and sixteenth centuries we can see, in notarial practices for identifying property sites, the slow unfolding of an increasingly standardized and official urban map based on streets. The development of this map gradually unhinged the linguistic and cognitive basis for alternative cartographic representations of possession and identity, a process more or less complete by the eighteenth century, when graphic maps

[40] E.g., ibid., fol. 34r.
[41] ADBR 3B 812, fol. 21r.
[42] Ibid., fol. 37r.

bearing the names of streets appeared for the first time. But the standardizing process was creating ripples in linguistic cartography long before this, for we can watch as episcopal and royal agents of record, influenced by this cartographic drift, gradually changed the language of space in their property records, abandoning islands in favor of streets by the sixteenth century. The overall process was part of a standardizing agenda not sponsored in any way by the Angevin state; it emerged first in the record-keeping habits of a relatively small group of notaries, usually numbering no more than thirty or thirty-five in a given year, and gradually spread from there into the cartographic languages of other agents of record. It was a cartographic transformation in linguistic form whose origins preceded the graphical *furor geographicus*. This transformation was in no way unique to Marseille. One can expect to find parallels in the record-keeping practices of notaries and other agents of record elsewhere in Europe, although the paths taken, surely, were different.

Transformation was initially limited to the description of property sites in notarial, seigneurial, and governmental records. The language of streets filtered more slowly into vernacular linguistic cartography, and did not immediately change how people constructed their addresses. Even in the sixteenth century, vicinities and landmarks continued to push through the increasingly standard notarial map. This can be explained in part by the fact that Marseille's late medieval notaries showed a marked disinterest in attaching addresses to individuals. The association between address and identity, it turns out, developed most precociously in records kept by seigneurial officials. The addresses in these records were the context-dependent addresses based on vicinity and landmark preferred by vernacular linguistic cartography. The notarial cartography of streets and the seigneurial association between identity and address were separate currents and followed different courses. At the confluence of these two streams in the nineteenth century, which is well beyond the purview of this book, there emerged a common language for mapping possession and identity, a language of streets, cities, regional entities, and eventually numeric rational postal codes that would, soon thereafter, become a major element in the identity categories used by the modern bureaucratic science of classification.

THE PUBLIC NOTARIATE

My claims have been large, and in them lies the thesis of this book. Of course, I am not alone in attaching a certain degree of political or epistemological agency to agents of record in medieval Europe. Medieval bureaucracies or administrative practices, whether civil or ecclesiastical, have always drawn their share of attention, some of it explicitly Weberian in focus.[43] In recent years, there has also been growing historical interest in the public notariate and what has been called notarial culture.[44] I will discuss notaries as cartographers in chapter two, but since the agency of the public notariate is a theme that runs through the entire book, it is important here to provide an introduction to the institution.

The genealogy of the notariate extends back to antiquity, and there was a notarial presence in Europe associated especially with the church throughout the early middle ages.[45] Before the twelfth century, charters written by scribes or notaries were inscribed on loose sheets of parchment, given over to the interested parties, and sometimes collected into cartularies—and sometimes not. Their survival was haphazard at best, and it is no surprise that the vast majority of existing cartularies come from ecclesiastical institutions. As the volume of notarial business grew in the twelfth and thirteenth centuries, notaries fell into the habit of copying the *précis* or protocol of each act into casebooks and then drafting instruments on the basis of these protocols. Protocols soon acquired a degree of legal force, and notarial dynasties became, in effect, the public archive. This marks the beginning of the system of public notaries.[46]

Owing to the demographic and commercial expansion and the political transformations of the high middle ages, the public notariate became a massive and growing presence in southern Europe from the twelfth century onward, and, a century or two later, in the north as well. The sheer scale of the institution cannot fail to impress. Over the middle of the fourteenth century, between 1337 and 1362, one hundred and sixty notaries plied their craft in Marseille, a city of around 25,000 inhabitants

[43] A particularly significant recent work is John W. Baldwin, *The Government of Philip Augustus: Foundations of French Royal Power in the Middle Ages* (Berkeley, 1986). See also Julius Kirshner, ed., *The Origins of the State in Italy, 1300–1600* (Chicago, 1995).

[44] I discuss the literature in more detail in chapter two. For a general discussion of the role of notaries in late medieval and Renaissance legal culture, see Lauro Martines, *Lawyers and Statecraft in Renaissance Florence* (Princeton, 1968), esp. 34–38. Peter Burke uses the term "notarial culture" in *The Historical Anthropology of Early Modern Italy: Essays on Perception and Communication* (Cambridge, 1987), 113, 128. My thanks to Ed Muir for bringing this to my attention.

[45] Rosamond McKitterick, *The Carolingians and the Written Word* (Cambridge, 1989), 115–26; Thomas F. X. Noble, "Literacy and the Papal Government in Late Antiquity and the Early Middle Ages," in *The Uses of Literacy in Early Mediaeval Europe*, ed. Rosamond McKitterick (Cambridge, 1990), 92–93.

[46] See John Pryor, *Business Contracts of Medieval Provence. Selected "Notulae" from the Cartulary of Giraud Amalric of Marseilles, 1248* (Toronto, 1981).

before the Black Death. Many worked for ecclesiastical and secular lords, copying out great rent registers, drafting letters, and compiling other documents demanded by their clients and employers. Others, no less than seventy-seven over this twenty-six year period, worked at least part of the time for the general public. These were distinguished by the title of "public notaries"; they wrote out contracts of many types in casebooks that they usually maintained on a year-to-year basis.[47] In Marseille, these casebooks were upward of one hundred folios in length, averaging around one contract per folio. Few have survived the ravages of time. Of the eight hundred that may have been produced in the twenty-six years between 1337 and 1362, only seventy-two have survived, containing around 6,600 acts.[48] In any given year in the mid-fourteenth century there were probably around thirty-five master notaries, assisted by their sons, sons-in-law, or other apprentice notaries, who were actively producing casebooks.[49] The proportion of notaries to the general public responded to demand, and was therefore influenced by population fluctuations, commercial activity, the market in property, the availability of credit, and a complex array of other factors.[50]

The public notaries of Marseille were peddlers of the law, sometimes working out of their homes or offices but more often circulating broadly throughout the city and the surrounding countryside. In these perambulations they met a lot of people. In his twelve surviving casebooks between the years 1348 and 1362, the notary Jacme Aycart conducted his trade in the presence of as many as 4,500 different people, both clients and witnesses.[51] The vast majority of these were residents of the city, either citizens or recent immigrants. A fifth of these were women. Collectively,

[47] The prosopographical index (see below, p. 38) identifies one hundred and sixty men as notaries, some of whom are known only because they show up incidentally in the surviving casebooks, perhaps as a neighbor or as a witness. I calculated the number of active public notaries (in this case, at least seventy-seven) by tracking those notaries who produced known notarial instruments or casebooks.

[48] On the question of survival see also Louis Stouff, "Les registres des notaires d'Arles (début XIVᵉ siècle–1460). Quelques problèmes posés par l'utilisation des archives notariales," *Provence historique* 100 (1975): 305–24.

[49] See AM FF 166, fols. 6v–11v. This cartulary of public proclamations lists thirty-three registered public notaries active in 1350. At least one notary active in the upper city, Esteve Venaissin, was not included in this list, suggesting that the register may have listed only those notaries active in the lower city. It is safe to assume that the total number of active notaries per year was thirty-five or more.

[50] For example, notarial activity increased significantly in the five years after the Black Death of 1348 to accommodate the vast increase in business arising from the tangled inheritances that fell to the survivors. See my "Accommodating Plague in Medieval Marseille," *Continuity and Change* 11 (1996): 11–41.

[51] This is the figure indicated by the prosopographical index.

Jacme's known clientele constituted at least a quarter of the total resident population of post-plague Marseille. As this indicates, by the mid-fourteenth century few people, certainly few propertied people, went through life without encountering the notaries in some official capacity on one or more occasions. When we consider all the surviving records between 1337 and 1362, not only notarial but also judicial, seigneurial, fiscal, and others, we find around 14,000 different men, women, and children who are mentioned at least once. The contact between people and notaries was staggering, in some respects rivaling the contact between the clergy and the Christian population. Despite this contact, notaries are not credited, as are the clergy, with a role in the shaping of western European culture and society. Even to place them in parallel categories clashes with our sensibility of what was important in the medieval world. But unlike the clergy, notaries did not write about themselves as figures of importance.

Like the clergy, however, the notaries and scribes of medieval Europe have left us an enormous treasure, thousands upon thousands of notarial casebooks and parchments piled up in great drifts in the medieval archives of regions, cities, and small towns all over the southern lands and, by the fifteenth century, in the north, as well. From the earliest survivals from twelfth-century Genoa, extant notarial casebooks increase rapidly in number over time, until by the early modern era there is an embarrassment of riches. From the decades between 1250 and 1500 there are 1,800 extant notarial casebooks concerning Marseille alone. Assuming a survival rate of one in ten, the original total was around 18,000.[52] The notariate of fourteenth-century Marseille, in turn, was puny in comparison to some Italian cities in the thirteenth century, notably Florence with its estimated 600 notaries, Pisa with 300, and Genoa with 200.[53] Although the hypothetical quantity of original casebooks has, to my knowledge, never been estimated, the public notariate of southern Europe surely produced hundreds of thousands of casebooks in the centuries before 1500. At the rate of around one hundred acts per casebook which was an average typical for Marseille, these casebooks collectively harbored tens of millions of acts. This incredible richness says a great deal about the transformations of the later middle ages. Each act, like a snapshot, took a legal moment,

---

[52] The survival rate in the middle of the fourteenth century was around one in fourteen (there are 2.3 surviving casebooks per year in mid-fourteenth-century Marseille, all that is left of the casebooks produced by thirty-five active notaries per year). For my purposes I have assumed the figure of one in ten because the survival rate from the fifteenth century was probably better.

[53] For the figures for the latter three cities see David Herlihy, *Pisa in the Early Renaissance: A Study of Urban Growth* (New Haven, 1958), 10–11.

a transaction, a set of facts—or, sometimes, fabrications—and froze the scene on paper, thereby recording it for a posterity that, in all likelihood, would never use it again. But the snapshot was there and served as a complement to memory as long as the people who had a stake in the event remembered where to find the document and were able to keep the ravages of fire, water, insects, and time at bay.[54]

The public notaries provided a huge range of legal contracts to a general public ever more interested in their services, thereby offering access not only to credit but also to testaments, dotal acts, legal services, compromises, commercial partnerships, commercial exchanges, property conveyances, apprenticeships, labor contracts, inventories, guardianships, emancipations, and so on. Such access was fundamentally egalitarian. The egalitarian nature of notarial activity, at least in Marseille, is not something I need to prove systematically with statistics. After all, it simply leaps out of the documentary record, and is something that will manifest itself in the chapters that follow. It suffices here to note that of the 2,227 men, women, and children mentioned in various clauses in the seven casebooks (extant between 1348 and 1362) of the notary Peire Aycart, no fewer than 485 (22 percent) of these people are identified either as laborers or fishermen, or the sons, daughters, wives, widows, or mothers of laborers or fishermen, the least favored professions in fourteenth-century Marseille.[55] Inexact identity clauses ensure that many more are unknown to us. We can be confident that notarial services were used with proportionally greater frequency by wealthier people, of course, and the opportunities offered by Peire Aycart and his notarial colleagues did not extend to slaves, lepers, and the very poor. Ruffians, drunkards, and others with little interest in the things notaries offered did not bother to employ them. Domestic servants did not often have the occasion to do so either, especially in an age when service contracts for domestic labor appear to have been exclusively oral. But virtually everyone else was capable of using, and did use, notarial services.

In the names of the interested parties, the public notaries recorded a wide range of obligations, debts, and rights in property. Many or all of these relationships and rights existed in some form in customary legal practices, and some were made the object of written acts before the advent of the public notariate in the twelfth century. These acts were not uncom-

<hr />

[54] As Mario Montorzi notes, citing Petrarch, writing is only as good as the memory that keeps track of what has been written and read. See *Fides in rem publicam: ambiguità e techniche del diritto comune* (Naples, 1984), 243–44.

[55] See ADBR 355E 34–36 and 355E 290–93.

mon in the early middle ages. As many have observed, the epistemological transformation of the later middle ages did not emerge from a simple
shift from memory to written record per se.[56] What is most significant, I
believe, is the massive increase in *scale* that was made possible in large part
by the egalitarian nature of notarial activity—and also by the demographic, economic, monetary, political, and legal-judicial transformations
of the high and later middle ages.

For my purposes, the massive increase in the sheer quantity of written
acts meant two things. First, public notaries—and other agents of record
for that matter—were writing more and more identity clauses, which is
to say, clauses in legal contracts that identified entities existing in the
real world: people, units of property, types and quantities of merchandise,
and sums of money. The greater frequency of identity clauses meant
that agents of record were creating more and more verbal descriptions
for something that people never had to think about a great deal before,
like the possible categories of identity. This resulted in a vastly larger
linguistic field in which linguistic conventions could subsequently
develop. Put differently, there was simply more talk about identity and
hence more effort to define it. As Harvey Sacks has argued, categorization plays an important role in conversation.[57] According to this argument, the larger conversational space developing over the later middle
ages helped create or refine languages of classification. One does not have
to postulate a guiding intellectual genius—a scientific revolution, a centralizing state, a set of legal norms developed in university settings—
to explain this change. Nor, I think, does one need to postulate any
revolutionary consequences resulting from a switch to written forms of
communication or print media.[58] The identity clauses that concern me
here would have emerged from oral conversations between agents of
record and clients. The specific technology of communication, here, is
ancillary to the more important transformation: far more conversations
about identity.

[56] Wendy Davies and Paul Fouracre, eds., *The Settlement of Disputes in Early Medieval Europe* (Cambridge, 1986); McKitterick, *Carolingians and the Written Word*; Patrick J. Geary, *Phantoms of Remembrance: Memory and Oblivion at the End of the First Millennium* (Princeton, 1995). See also Clanchy, *From Memory*; Brian Stock, *The Implications of Literacy: Written Language and Models of Interpretation in the Eleventh and Twelfth Centuries* (Princeton, 1983).

[57] Harvey Sacks, *Lectures on Conversation*, ed. Gail Jefferson, 2 vols. (Oxford, 1992), 1:40–48. See also the index, s.v. "categories and classes."

[58] Jack Goody, *The Logic of Writing and the Organization of Society* (Cambridge, 1986); idem, *The Interface between the Written and the Oral* (Cambridge, 1987); Elizabeth Eisenstein, *The Printing Press as an Agent of Change: Communications and Cultural Transformations in Early Modern Europe*, 2 vols. (Cambridge, 1979).

Second, the things identified by such clauses were increasingly numerous, obscure, and hard to know. The egalitarian nature of contractual behavior meant that ever greater numbers of people, units of property, and other entities were being identified. It is easy to know a lord. It is less easy to know a laborer. Perhaps it is not surprising to learn that laborers are more carefully identified than noblemen in notarial contracts from medieval Marseille. The same is true for property. Throughout most regions of Europe in the high and later middle ages there was a growing market in property. Marseille was no exception to this general rule.[59] The market in property brought attention not only to well-known castles or great seigneurial estates but also to relatively tiny and insignificant units of property. The resulting property conveyances described a relationship that increasingly implicated members of different communities of knowledge, and the lack of shared knowledge encouraged the formation of some kind of institution that would mediate between these different communities: first, the public notaries; later centralized state archives.

This is a point worth elaborating even in a highly theoretical and speculative way. Medieval archives were first and foremost the archives of social memory.[60] Throughout the early middle ages these archives were perfectly adequate to the task of knowing who was who and what was what. In a relatively thinly populated world, for example, it is easy to "know" a castle or manor. These entities exist in the public domain when the community of knowledge is the entire local community. It is easy to know any portion of a lord's estate, even when that estate is geographically fragmented, because the lord, either in person or through proxies, periodically pays muscular and public visits to the outlying portions of the estate. Only with the changing demographic conditions associated with the twelfth century—especially the emergence of densely populated, demographically complex cities—do we begin to see changes in the social construction of community knowledge. It is not easy to "know" a house located in the large, complex cities that emerged in the twelfth century and after. Neighbors will know a house is there at a particular location. They will watch the comings and goings next door and will distinguish between servants and proprietors. They will try to witness any property conveyances

[59] For Marseille, see my "Accommodating Plague," 22–31.
[60] Useful works in this field include Maurice Halbwachs, *On Collective Memory*, ed. and trans. Lewis A. Coser (Chicago, 1992); Peter L. Berger and Thomas Luckmann, *The Social Construction of Reality: A Treatise in the Sociology of Knowledge* (Garden City, N.Y., 1966); Eric Hobsbawn and Terence Ranger, eds., *The Invention of Tradition* (Cambridge, 1983); James Fentress and Chris Wickham, *Social Memory* (Oxford, 1992); Pierre Nora, ed., *Realms of Memory: Rethinking the French Past*, trans. Arthur Goldhammer (New York, 1996).

involving the house and will notice who pays rent when agents of the direct lord come knocking on the door. Similarly, they keep track of the living and the dead. In all these ways they keep track of ownership. Likewise, members of a descent group—those who have an interest in the estates belonging to their relatives—tend to make a point of knowing just what that estate consists of. But in a large city like Marseille there were many such houses, and knowledge of them—the who, what, and where of possession—did not necessarily circulate outside of the communities of knowledge particular to kinfolk and neighbors. What we see here is a progressive fragmentation of communities of knowledge in urban settings during the later middle ages.

This had an immediate relevance for contractual behavior. It is relatively easy for exchanges to take place and contracts to form within communities of knowledge, because reputations for honesty are known to all members of the community, and anyone caught violating the norms of reciprocity or the terms of a contract can suffer group sanctions. But in the complex settings of late medieval urban communities and the late medieval economy, with its frequent exchanges across fragmented communities of knowledge, what serves to provide guarantees on behavior?

This, in essence, was the function of the public notariate. Notarial contracts were backed up by rapidly developing courts of law. Since notaries provided access to a reasonably efficient form of coercive power, their presence not only allowed for a more egalitarian access to contractual activity—the weak, notably Jews, women, and children, could participate with relatively few constraints—but also for greater impersonality, greater social and physical distance, in the relationships between clients. Long-distance commercial agreements; loans involving Christians and Jews; labor contracts between nobles and workers; testaments governing property relationships between the living and the dead; and procurations giving a litigant access to an expert legal pleader illustrate the kinds of social, temporal, and physical distancing enabled by notarial activity. This distancing placed the relationship on an abstract footing and was a necessary corollary not only to the commercial revolution but also to the fragmentation of communities of knowledge. In a similar way, the growing market in property, facilitated if not created *ex nihilo* by notarial culture, made the relationship between people and property an increasingly abstract one. The use of Roman-canon law in medieval Europe made it easier to imagine possession without periodic manifestations of physical presence; it complemented the existing trend in medieval Europe that turned property into a commodity. Egalitarianism and distancing created contractual relationships that existed across communities of knowledge. Thus, as is often said, notaries were and are professional witnesses, and their written instruments, in the eyes of contemporary observers, bound the truth more efficiently than did fallible human memory.[61] Since their contracts facilitated exchanges across communities of knowledge, it fell upon the notaries to guarantee the authenticity of the exchange.

A science of identification and classification naturally evolved out of this emerging need to identify the obscure. This is true even though the identifications developed and used by late medieval notaries were, by modern standards, none too precise. They did not need to be, because notarial acts did not, as of yet, pretend to identify entities expertly, officially, and scientifically. They did not pretend to be the impersonal archives created by modern state bureaucracies. We must assume that notaries knew many of their clients personally and used mnemonic devices to remind them of the people they did not know so well. The notary's prodigious memory and status as an official witness alone was sufficient to guarantee the authenticity of his identifications. As a result, there was no immediate legal pressure on notaries to craft ever more elaborate identity clauses. In the same way, and for the same reasons, there was no immediate pressure on governmental agencies to replace social memory with written archives. Although a bureaucratic science of identity classification did develop across the later middle ages, it developed slowly and fitfully, and the nature of this change denies the presence of a guiding intellect.

This is so because of the political nature of the notariate, both autonomous and decentralized. Public notaries worked for the general public, not for agencies of state, even though they could and often did take up temporary judicial or curial offices or even private posts working for ecclesiastical and secular lords. The contracts they drew up as public notaries were made authentic simply by the fact of being notarized. In principle, notaries theoretically acquired this power of authentication from a sovereign authority, granted to them after a period of training and apprenticeship. For example, in Marseille the authority was the count of Provence. Notaries authenticated their contracts by means of a signature or seal, and yet although they were licensed by an authority, such notaries still cannot be identified as members of a rational-legal bureaucracy responsive to state concerns even if they had certain public responsibilities, chief among them the requirement to safeguard their casebooks and

---

[61] The capacity of the public notary to hold *publica fides* is nicely summarized in Armando Petrucci, *Writers and Readers in Medieval Italy: Studies in the History of Written Culture*, ed. and trans. Charles M. Radding (New Haven, 1995), 152–57, 240–45.

pass them down within securely identified notarial dynasties.[62] Most notarial casebooks in Provence and France, in fact, were passed from notary to notary down through the ages until the late nineteenth century and early twentieth centuries when the government began to invite the current holders to deposit their casebooks in state archives.

In a sense, these notarial lineages were understood to constitute a kind of public archive over which local governments attempted to exert some control. However, the archive was, to coin a term, a "dispersed archive" heavily suffused by the techniques of memory and caught between public and private interests. The format of the dispersed archive is quite at odds with the centralized archives typical of modern states.[63] Again, there was a tight link between notaries and courts of law, but this does not mean that notaries were in any sense appendages of the courts, or even governmental officials, in any sense of the word. Consequently, for all these reasons, there were no clear avenues for the exercise of state or sovereign influence over those aspects of notarial activity that will concern us here.

Most important is that sovereign or state involvement in notarial activity did not extend to the drafting of the language of acts. This fell into the purview of the *ius commune*, the legal system in which notaries were trained. The *ius commune* was an international body of law discussed in university settings, particularly at Bologna, and theorized by an international body of professional jurists. It was not an organ of state and can be seen as antithetical to state interests.[64] Within the guiding framework of the *ius commune*, individual notarial communities, such as the notariate of Marseille, had considerable latitude to develop their own legal customs on matters that did not seem of compelling importance, such as identity clauses and plat descriptions. Thirteenth-century notarial formularies such as those of Bencivenne and Salatiele are untouched by cartographic speculation. Both provide samples of property conveyances concerning urban property—as it happens, they use the geographical designation of the parish (*parrochia*) and not the street—but neither pauses to note just how this clause should be crafted.[65] Thus, although the widespread adop-

---

[62] This dynastic awareness, naturally, promoted patriarchal attitudes on the part of notaries; see Julie Hardwick, *The Practice of Patriarchy: Gender and the Politics of Household Authority in Early Modern France* (Pennsylvania State University Press, 1998).

[63] Some of the contradictions inherent in the duty of notaries to archive are discussed in Ezra N. Suleiman, *Private Power and Centralization in France: The Notaires and the State* (Princeton, 1987), 38–47.

[64] Manlio Bellomo, *The Common Legal Past of Europe, 1000–1800*, trans. Lydia G. Cochrane (Washington, D.C., 1995).

[65] See Salatiele, *Ars notarie*, ed. Gianfranco Orlandelli, 2 vols. (Milan, 1961), 2:229, and Bencivenne, *Ars notarie*, ed. Giovanni Bronzino (Bologna, 1965), 38. Similarly, they do not provide any reflections on addresses in identity clauses.

tion of the basic principles of Roman-canon law may have created pressure among the public notariate to develop a system of classification, there were no existing guidelines on what the categories of identity should consist of. These categories emerged slowly out of practical experience. Of significance is that they emerged at all, pushed neither by interests of state nor by a scientific revolution in legal cartography.

THE BUREAUCRATIC SCIENCE OF CLASSIFICATION

Why should this interest us? To appreciate the significance of this epistemological transformation, we need to consider another major theme of this book, namely, the nature and evolution of record-keeping bureaucracies or institutions.[66] Medieval notaries did not constitute a bureaucracy, at least in any conventional or Weberian sense of the word, since the term bureaucracy implies an association with the interests of modern states.[67] Yet the notariate was an institution that was slowly acquiring techniques whereby to classify the identities of people and property. Other record-keeping institutions in medieval Europe were engaged in a similar endeavor, and the public notariate is especially interesting not because it was unique but simply because it was so massive. The techniques of classification that emerged in the practices of notaries and other agents of record in late medieval Europe were available for use in later schemes of bureaucratic classification.

Certain state bureaucracies, namely, those whose institutional function requires them to keep track of individuals or their possessions—notably houses, other landed property, automobiles, and professional degrees— have a penchant for classifying their subject matter. In so doing, they participate in what James C. Scott calls a "state project of legibility and simplification."[68] Historically, the urge to classify arises to some extent whenever record-keeping becomes centralized and the number of records becomes inconveniently large for the institution to regulate easily—

---

[66] The literature on bureaucracy is vast. The general theoretical works I consulted include Robert K. Merton et al., eds., *Reader in Bureaucracy* (New York, 1952); Michael Crozier, *Le phénomène bureaucratique* (Paris, 1963); Michael T. Herzfeld, *The Social Production of Indifference* (Chicago, 1992); Bernard S. Silberman, *Cages of Reason: The Rise of the Rational State in France, Japan, the United States, and Great Britain* (Chicago, 1993); Giovanni Busino, *Les théories de la bureaucratie* (Paris, 1993).

[67] E. Suleiman depicts the nineteenth- and twentieth-century French notariate as an ambiguous (and singularly non-Weberian) bureaucracy; see *Private Power and Centralization in France*, 299–330 and passim.

[68] James C. Scott, *Seeing Like a State: How Certain Schemes to Improve the Human Condition Have Failed* (New Haven, 1998), 9.

among other things, beyond the capacity of the memory of the recorder. More significantly, the urge to classify arises when record-keeping bureaucracies become aware of a need to identify individuals or possessions according to abstract or universal categories of thought that transcend time and place.[69] In the literature on this subject, the need to classify can be seen as an instrument of state-building. As Bernard S. Cohn and Nicholas B. Dirks remark, "The legitimizing of the nation state proceeds...by constant reiteration of its power through what have become accepted as natural (rational and normal) state functions, of certifying, counting, reporting, registering, classifying, and identifying."[70] Scott phrases the role of state interest in classification in these terms:

> How did the state gradually get a handle on its subjects and their environment?... [P]rocesses as disparate as the creation of permanent last names, the standardization of weights and measures, the establishment of cadastral surveys and population registers, the invention of freehold tenure, the standardization of legal discourse, the design of cities, and the organization of transportation [seem] comprehensible as attempts at legibility and simplification. In each case, officials took exceptionally complex, illegible, and local social practices, such as land tenure customs or naming customs, and created a standard grid whereby it could be centrally recorded and monitored.[71]

This impulse predates the high modernism that is the focus of Scott's study. As Scott himself notes, in the sixteenth century and beyond, record-keeping bureaucracies that were attached to incipient states began to expand considerably in western Europe and ranged from census and revenue bureaus to parochial offices and the criminal justice system.[72] By

---

[69] Gérard Noiriel, *Le creuset français: histoire de l'immigration XIXᵉ–XXᵉ siècles* (Paris, 1988), 71–124, 353; idem, *La tyrannie du nation: le droit d'asile en Europe 1793–1993* (Paris, 1991), 155–180; Ian Hacking, "Making Up People," in *Reconstructing Individualism: Autonomy, Individuality, and the Self in Western Thought*, ed. Thomas C. Heller, Morton Sosna, and David W. Wellbery (Stanford, 1986), 222–36. Also relevant is Carlo Ginzburg, "Clues: Roots of a Evidential Paradigm," in idem, *Clues, Myth, and Historical Method*, trans. John Tedeschi and Anne C. Tedeschi (Baltimore, 1989), 96–125.

[70] Bernard S. Cohn and Nicholas B. Dirks, "Beyond the Fringe: The Nation State, Colonialism, and the Technologies of Power," *Journal of Historical Sociology* 1 (1988): 225.

[71] Scott, *Seeing Like a State*, 2.

[72] Ibid., 3. For France, see also Alexis de Tocqueville, *The Old Régime and the French Revolution*, trans. Stuart Gilbert (Garden City, N.Y., 1955); Roland Mousnier, *The Institutions of France under the Absolute Monarchy, 1598–1789*, trans. Brian Pearce (Chicago, 1979); Ezra N. Suleiman, *Politics, Power, and Bureaucracy in France: The Administrative Elite* (Princeton, 1974).

the nineteenth century, they had become massive institutions, and their demands fostered the refinement of personal identity templates. To take a particularly telling example, the growing use of excommunication by the Catholic church from the thirteenth century onward encouraged the development of registers describing the state of the souls in given parishes. The earliest extant register in France, from the diocese of Narbonne in the year 1404, gives only the number of communicants per parish. By the early seventeenth century, ecclesiastical guidelines were demanding the surname, forename, sex, age, and spiritual state of each individual parishioner. By 1706, the template also included an address defined in terms of quarter and street. All the official need do was to write in the necessary places in the spaces provided; the template itself was already fixed.[73]

As in cases like this one, early modern bureaucracies were seeking to replace context-dependent identities, or social memory, with transcendent identity categories. In the record-keeping practices of ecclesiastical as well as state bureaucracies, the emerging templates of personal identity were made up of several possible categories, and the number of categories by which a person could be defined grew over time. In early twentieth-century France, the Ministère de l'Intérieur printed up forms for use in identifying suspects arrested by the police. The "notice individuelle" used the following categories: surname, forenames, birthdate, birthplace, domicile, parentage, parent's profession and domicile, profession of accused, previous domicile, marital status, financial status, level of education, military service, and even, in a separate part of the form, an echo of the long-dead inquiry into the state of souls: "information on morality and reputation."[74] The templates developed in other political jurisdictions might use any of these categories along with race, ethnicity, religion, and sex, depending on local circumstances and other factors. In certain records, especially those concerned with identifying potentially recidivist criminals without fixed domiciles, record-keeping bureaucracies from the late nineteenth-century onward began to rely heavily on anthropometric features such as height, hair color, chest size, shape of head, face, nose, ear, lips, eyelid, and eyebrow, eye color, skin color, body size, and so on, captured by verbal descriptions and sometimes accompanied

---

[73] G. Couton and H.-J. Martin, "Une source d'histoire sociale: le registre de l'état des âmes," *Revue d'histoire économique et sociale* 45 (1967): 244–53.

[74] See, for example, the records of arrest between 1907 and 1940 found in ADBR series 4M 281 to 514; the individual dossiers of suspects include these "notices individuelles." The template changed slightly from decade to decade.

by a photograph or fingerprints. Such techniques were associated, in France, with the late-nineteenth-century criminal anthropologist Alphonse Bertillon.[75] Some of these categories were used in the later middle ages, although they were not made the subject of a sociological or anthropometric science. The difference is one of both scale and systematic application. By the late nineteenth century there were many more categories of identity in common use, and the categories used by any template were applied systematically to the description of the individual in question, in conformity with a pre-established template.

Each category of any identity template, past or present, assumes that human identity will vary across a predefined set of spectra. The role of each label is to situate a unique individual on those spectra, in much the same way that a cadastral map or a plat situates property on the landscape. Hence, a birthdate defines an individual on a spectrum of possible ages; race, on a spectrum of possible human races; and nose according to several categories of possible noses: rectilinear, sinuous, and so on. An address, similarly, situates the individual within geographic space. The categories assume that all individuals will fit somewhere on the specified spectrum, leading to such things as the long and rather absurd list of possible professions on the U. S. census. People who do not fit, such as mixed-race individuals and hermaphrodites and those with indefinable occupations, pose problems for the templates created by record-keeping bureaucracies. In the category of address the homeless do not quite fit the demands of the template, and geographical mobility, although less sinister, places strain on the ability of bureaucracies to track individuals. In early modern France, parish officials in charge of keeping the registers of the state of the souls were required to track mobile individuals.[76] In late nineteenth-century France, more systematic strategies for identifying foreigners and *nomades* began to emerge, resulting, in 1912, in a law imposing identity cards on this category of individual. The identity card consisted of the anthropometric categories of identity established by Bertillon; duplicate copies were kept in police archives.[77]

We must assume that the categories chosen are important and meaningful in some way, quite apart from the utility of anthropometry in criminological contexts. Identity categories do not often include income, or primary language, or intelligence, or charisma, or fashion sense, or self-

---

[75] See Alphonse Bertillon, *Identification anthropométrique*, 2 vols. (Melun, 1893); see also Jean-Marc Berlière, *Le monde des polices en France: XIXᵉ–XXᵉ siècles* (Brussels, 1996), 41–48.
[76] Couton and Martin, "Registre," 245.
[77] Noiriel, *Tyrannie*, 176–77.

confidence, or drive, or other things that individuals might think of as being important categories of their identity. In a democracy, these things are not supposed to be significant. Certain medieval epithets, like Charles the Simple, Thord the Chatterer, Louis the Pious, or Charles the Bold, provide by way of example a nice counterpart to this modern practice. Race, age, sex, address, profession, and personal features are aspects of identity that in all their permutations are, for whatever reason, understood to be impersonal and transcendent and therefore more valid categories of identity than degree of simplicity, chattiness, piety, or bravery. Consequently, the history of identity clauses and identity templates will not necessarily tell us what people thought of themselves. It will only tell us how people defined themselves, or were asked to define themselves, in certain bureaucratic contexts.

These bureaucratic contexts are shaped by political culture and common usage, and so the history of the identity templates used by record-keeping bureaucracies is something more than a mere history of bureaucratic science. Some of the categories typical of the bureaucratic templates that had emerged by the nineteenth century were novel. Birthdate, for example, was not often used before the sixteenth century in Europe. Sex as a separate category was never used in the records from fourteenth-century Marseille. As Robert Bartlett observes, other categories, such as race or ethnicity, customs, language, and law, were relatively common in medieval Europe, especially wherever mixed populations could be found.[78] Some categories of identity typical of premodern bureaucratic regimes, such as parentage or lineage and noble status, were progressively dropped by some modern bureaucracies, although different countries did so at different rates.

The modern idea that any citizen-subject can be identified according to a bureaucratic template that is universal within the boundaries of a given state serves to homogenize diverse populations. Identity itself can be part of a hegemonic discourse, and, as Benedict Anderson has observed, imagined communities of all descriptions have a vested interest in framing the identities of subjects and citizens.[79] Record-

---

[78] Robert Bartlett, *The Making of Europe: Conquest, Colonization and Cultural Change, 950–1350* (Princeton, 1993), 197–98.

[79] Benedict Anderson, *Imagined Communities: Reflections on the Origin and Spread of Nationalism* (London, 1983); see also idem, "Census, Map, Museum," in *Becoming National: A Reader*, ed. Geoff Eley and Ronald Grigor Suny (New York, 1996): 243–48. Modern nation-states, able to harness the identity-shaping power of print capitalism and nationalized systems of primary education, are particularly efficient in this effort, but the general idea would hold for any imagined community.

keeping bureaucracies play a large role in this identity-framing process simply because paperwork requires a system of classification. Situating individuals across the various spectra acknowledges human diversity but nonetheless underscores a common humanity. This is a very democratic idea, and, as Gérard Noiriel argues, it probably contributed a great deal to the progressive interiorization of national identity and the creation of the imagined community of the nation-state.[80] After all, one participates in this democratic and national exercise every time one fills out a form.

The massive contact between the categories of analysis and identity created by record-keeping bureaucracies and the general public ensures the transmission of these ideas. In this bureaucratic relation there is, to extend Michael Herzfeld's comparison between the secular and the sacred, a secular counterpart to Christian pastoral care.[81] By the same token, it is also likely that some of the administrative categories created identities where none existed before. As Ian Hacking argues, "the sheer proliferation of labels . . . during the nineteenth century may have engendered vastly more kinds of people than the world had ever known before." And further:

> The claim of dynamic nominalism is not that there was a kind of person who came increasingly to be recognized by bureaucrats or by students of human nature but rather that a kind of person came into being at the same time as the kind itself was being invented. In some cases, that is, our classifications and our classes conspire to emerge hand in hand, each egging the other on.[82]

Mary Douglas remarks in turn:

> The responsiveness to new labels suggests extraordinary readiness to fall into new slots and to let selfhood be redefined. This is not like the naming that . . . creates a particular version of the world by picking out certain sorts of things. . . . It is a much more dynamic process by which new names are uttered and forthwith new creatures corresponding to them emerge. Hacking's point is that people are not merely re-labeled and newly made prominent, still behaving as they would behave whether so labeled or not. The new people behave differently than they ever did before.[83]

---

[80] Noiriel, *Tyrannie*, 312–22.
[81] Herzfeld, *Social Production*.
[82] Hacking, "Making Up People," 226, 228.
[83] Douglas, *How Institutions Think*, 100.

These questions are particularly relevant to identity categories such as madness, sexual orientation, race, class identity, and gender, but the principle can be generalized to all elements found in bureaucratic identity templates. To explore these issues we need a more systematic history of the identity clauses and identity templates developed and used by bureaucratic regimes, both past and present. Such a history is, or would be, an element of a larger sociology of identity that has been around at least since the time of Émile Durkheim and Max Weber.[84] Durkheim was especially attuned to the importance of classification to human society; more recently, anthropologists and sociologists such as Mary Douglas and Luc Boltanski have taken up the baton.[85] The history of the bureaucratic identity clauses, however, is surprisingly undeveloped and is an obvious subject for historical and serial analysis.[86]

Of all the categories of personal identity the most curious, and the least studied, is the address or legal domicile. The oversight is understandable, for the address has none of the innate appeal of race, gender, madness, or sexuality. But there are good reasons for studying the historical construction of the address. All the categories found on modern bureaucratic identity templates may be social constructs to some degree, race particularly so, but the address is the least intrinsic and least intuitive feature of a subject's identity. Among other things, it is the only major category of identity that can be changed easily and at will. Unlike the surname of a newly married woman, it is not even necessary, in the United States, to fill out a form notifying governmental agencies of the change, although if you want to receive mail it is convenient to do so. The possible responses, moreover, do not vary across a simple spectrum, like age or sex, and cannot be encompassed by an authoritative list, like race or profession. They are the only commonly used markers of identity (other than anthropometry in the past and the handwritten signature or DNA profile in the present) that are unique to individuals or, in the case of addresses, at least to families. New addresses are added by the thousands every year, and hence the address template must be a uniquely flexible form, defined by a nesting set of cellular, geopolitical entities culminating, almost invariably, in a street or its equivalent. Plats and streets can be added to the map at will. The new address is professionally surveyed and registered with appropriate governmental agencies and is subsequently recognized

---

[84] See the discussion in Douglas, *How Institutions Think*, 93–99.
[85] Luc Boltanski, *Les cadres. La formation d'un groupe social* (Paris, 1982); see also *L'enquête sur les catégories: de Durkheim à Sacks* (Paris, 1994).
[86] Anderson discusses some of the existing literature in "Census, Map."

and mapped by the postal service—in the United States a parastatal institution, elsewhere a unit of the state.

Despite the slipperiness of the address as a concept, addresses are favored in the classificatory schemes of modern record-keeping bureaucracies because they make it possible to find people and because they help denote unique individuals. In late nineteenth-century France, identity cards with anthropometric categories came into use precisely for those categories of individuals, namely foreigners and vagabonds, who had no fixed domicile. But even more than this, bureaucratic classifications create identities that are cellular and assemblable, identities that are useful for the nation-building project. The historical invention of the address as a nesting set of geopolitical terms was arguably one of the major geographic contributions to modern national identity, the other being the better-known creation of an idea of nation as a geographical entity with borders.[87] The two projects complemented one another, for the nation as a composite of the cellular identities of citizen-subjects, all of whom are neatly identified by means of a universal and modular address template, mirrored the nation itself as a cell within the world body. So important has the address become that few of us nowadays, with the possible exception of census takers whose job is made difficult by the homeless population, pause to consider the modern association between identity and address and the ways in which having an address frames us as citizen-subjects embedded within a series of geopolitical entities culminating in the nation-state.

MARSEILLE: A CASE STUDY

This book addresses some of the historical roots of this transformation. The terrain for such a study can either be broad and comparative or, as in the case here, microhistorical. Apart from the fact that the comparative evidence in the secondary sources is little developed, a case study can also offer a more detailed and close reading of the diverse array of available sources. In this book I will be using evidence from the massive notarial, judicial, seigneurial, and fiscal archives of late medieval Marseille, primarily in the mid-fourteenth century, but ranging also between the mid-thirteenth century, when the earliest of these records is found, to the mid-sixteenth century, when graphic maps of urban vistas emerge with sudden rapidity in Marseille and elsewhere in western Europe. Close

---

[87] In addition to works cited above, see Paul Alliès, *L'invention du territoire* (Grenoble, 1980).

familiarity with the documents from a short span of time, in this case a quarter century between 1337 and 1362, has allowed me to build up an intimate acquaintance with both the geographic sites of the city and with the people who sought to describe them. I have indexed most of the extant records for this period and, by virtue of record linkages and careful prosopographical (or profile) study of individuals, have created a database of some fourteen thousand men, women, and children encountered at least once in the extant records, allowing me to track down individuals and compare addresses with relative ease. I will refer to this, from time to time, as the "prosopographical index." The index, naturally, has a bias toward men and women who appear more frequently in notarial and rent registers, and thus toward property-owning individuals. Property-ownership in Marseille, however, extended deep into the ranks of the middling and poorer sorts. Rent registers kept by ecclesiastical and secular lords in Marseille list several thousand proprietors of urban and rural property, among whom there are 770 whose profession or status is given. Of this number, 246, or 32 percent, were agricultural laborers and 45, or 6 percent, were fishermen; these were among the professions with the least status.[88] The egalitarian nature of notarial activity also ensures that a fairly broad array of social groups is represented in the index. Each record in the index, the product of record linkages made between all available sources, includes some permutation of the following items: name, trade or status, known or probable place of residence, source(s), and miscellaneous information on kinship and other relevant matter.[89] This index, in many respects, is a source in itself. As the product of record linkages, it contains information of a different order than the information contained in the individual sources.

In the chapters that follow, I begin by sketching the city of Marseille, a city made interesting by its late medieval decay. Once a Mediterranean entrepôt of considerable stature in the twelfth and early thirteenth centuries, the city's commercial importance declined somewhat in the economic depression of the fourteenth century. A series of political setbacks over the fourteenth and fifteenth centuries, including the sack of Marseille in 1423 by the Catalans and the rise of Avignon and Aix-en-

---

[88] See ADBR B 538, B 1940, B 1941, B 1942, 5G 112, 5G 114, 5G 115, 5G 116, 1HD H3, 1HD B102, 4HD B1, 5HD B5, 5HD B6, 6G 485, and 355E 3.

[89] The quality of the index depends to a large extent on the care with which it was put together; see appendix 2. Here, it is worth remarking that the people of Marseille had fixed surnames; as a result, making linkages is much easier. On the issue of record linkages see David Herlihy, "Problems of Record Linkages in Tuscan Fiscal Records of the Fifteenth Century," in *Identifying People in the Past*, ed. E. A. Wrigley (London, 1973), 41–56.

Provence as centers of culture and government, kept the city a backwater. It was a mixed blessing, for the decline of potential revenues made the city less relevant in the international game of state-building. As a result, Marseille was more or less left alone by its Angevin overlords during the fourteenth and fifteenth centuries. Consequently, Marseille became a city with a well-developed culture of documentation that was to an unusual extent *not* shaped by an active state with a universalizing agenda, as one finds in Florence, Venice, or in the kingdoms of northern Europe. Of significance is that the records from late medieval Marseille reveal important changes in the ways in which urban maps and identities were drawn, and if we cannot explain these changes so easily in terms of state interest, then we are left with the possibility that the impetus for change either drifted in through international pathways of communication or, as argued here, developed independently.

With the ongoing development of notarial culture, notaries became cartographers, keepers of the most authoritative linguistic map of late medieval Marseille—at least the map that conforms most closely to modern, street-based expectations. This is true in a trivial way, since notaries wrote most of the documents pertaining to rights in property that have survived from the period. Yet as argued in the second chapter, the public notaries, working for clients ranging from common men and women to great landlords, acted unconsciously as map makers or archivists of the landscape, recording toponyms, locating houses and buildings, and identifying streets, plazas, markets, and squares. This cartographic role emerged because the legal requirements of the acts drawn up by notaries demanded that real estate be mapped and that the acts themselves be validated by reference to a time and a place. Over time, the notaries developed a distinct cartographic convention based on the perception of the city as a tracery of streets. This cartography was necessarily a legal one, but it was legal for less than obvious reasons. The acts that notaries drew up in Angevin Marseille had legal force not just because they conformed to a Roman legal norm concerning the act of mapping property sites but also because the notary himself was considered an agent of custom, an authority on certain matters of common knowledge such as questions concerning the map of the city. To us, the property sites found in house sales and similar acts *look* exceedingly imprecise. But they were not, as long as the notary who drew up the act held a map of the city in his head and knew where the property in question was located.

Marseille's notaries also served clients other than the general populace and in so doing they used different cartographies. In chapter three, I discuss parts of the city where, in the mid-fourteenth century, the street template favored by the public notaries spectacularly failed, namely, in the upper or episcopal city, an area dominated by Marseille's bishop and the insular template of the episcopal chancery. So powerful was this template in Marseille's upper city that many streets did not even acquire names until the sixteenth century. Public notaries, who preferred to think in terms of streets, often had to employ extraordinary circumlocutions to identify plats.

Chapter four addresses the templates typically used by artisans, retailers, laborers, and members of other occupational groups, both men and women. A single record that conveys a sense of vernacular cartographic discourse—kept in Provençal and therefore unmediated by the Latinate culture of the notariate—reveals that these people had a notable preference for the templates of vicinity and landmark. The considerable gap between vernacular and notarial cartographic languages reveals that a significant act of translation occurred during the cartographic conversations that took place between notaries and their clients whenever property changed hands. In these conversations, notaries took the Provençal nomenclature offered to them by their clients and made it conform, wherever possible, to the Latinate notarial norm.

Over time, the street became accepted by all as the official cartographic template. For artisans, retailers, laborers, and other ordinary speakers of the vernacular, this meant abandoning an unmappable social construct for a mappable, architectonic construct, at least in official discourse. "Social" addresses are still common. People in New York City will use entities like "Soho" or "the Village" in conversational identity constructs, avoiding the streets that are irrelevant to the status being claimed. Nonetheless, in the later middle ages, the universalizing trend in the mapping of property was spilling over into the processes whereby identity was mapped. Chapter five broaches this subject by exploring how agents of record constructed identity clauses in the fourteenth century. In notarial records, use of the address in identity clauses was fitful at best. Used rarely by nobles and Jews, it can be found slightly more often in the clauses identifying non-noble members of the free Christian population. Addresses were more commonly used in clauses identifying debtors, indicating that the interests of creditors could shape the content of identity clauses. They were used most commonly by officials who worked on behalf of great property lords, most notably the bishop and the cathedral canons but also a host of minor secular property lords. The need to address clearly emerged in the context of monetary obligations, and thus the late medieval culture of debt was a significant agent in creating an association between identity and address. This financial interest, however, does not explain why artisans, retailers, professionals, merchants, and members of the laboring population evince a marked tendency to use addresses; I will argue that these people, lacking the lineal consciousness of nobles and Jews, autonomously developed the custom of creating identities by means of reference to the landscape. The particular form of the address that came to be preferred by modern western bureaucratic practice, the street address, emerged when these existing customs of addressing gradually blended in the eighteenth century and beyond with the architectonic, street-based cartography already being developed by the public notaries.

As students of nationalism such as Peter Sahlins have been arguing, boundaries and frontiers do not simply exist; they are ideas created and fostered in the context of state-building and in the service of specific political goals. Sahlins's particular insight has been to show some of the local roots for this ideological production, thereby questioning the link between the interests of the state and the geographical awareness that emerged during this period. This book is an effort to describe the cartographic imagination of a society well before the state was a major player in the production of cartographic knowledge. By taking state interest out of our interpretations of how cartographic science developed in the language and thought of the later middle ages, it becomes possible to locate agency in decentralized record-keeping bureaucracies and explore, thereby, the local roots of ideological transformations.

CHAPTER ONE

Marseille

Marseille, the oldest city within the confines of modern France, is situated about thirty miles east of the mouths of the Rhone, on a short stretch of Mediterranean coast that runs roughly north-south. The old medieval port, now a port for pleasure craft, cuts eastward into a small gap between two hills, separated from the sea by the narrow mouth across which a protective chain used to be slung. The medieval city occupied only the northern shore of the small harbor, tracing an irregular triangle with the port at its base, the sea to the northwest, and the suburbs to the east. The site nestles in a basin ringed by high hills of limestone, an ancient sea bed pushed up and folded by the inexorable northward drift of crustal plates. Owing in part to these hills, medieval Marseille found it difficult to stamp its authority over a large agricultural hinterland in order to become a city-state like the other Mediterranean cities of Florence, Venice, Bologna, even Toulouse. Like Genoa, Marseille has always looked to the sea.

The city has had several pasts. One is the civic history of urban primacy and Greek origins; this can be found in the promotional literature of tourist guides, in general histories such as those of Raoul Busquet, Édouard Baratier, Pierre Guiral, and others, and in the work of archeologists interested in the Greco-Roman past.[1] Another is the progressive history of mercantile development and political liberation, characterized

---

[1] Édouard Baratier, *Histoire de Marseille* (Toulouse, 1973); Raoul Busquet, *Histoire de Marseille*, ed. Pierre Guiral (Paris, 1978).

most obviously by the work of Victor L. Bourrilly.[2] A third is the nostalgic history of local historians and amateurs like Bruno Roberty and Philippe Mabilly,[3] and a last, perhaps, is the Annales-inspired total history approach, found in the great *Histoire du commerce de Marseille* of the mid-twentieth century.[4] What follows is an accounting of all of Marseille's histories.

SOCIAL TOPOGRAPHY AND POLITICAL STRUCTURES

With a population estimated at 25,000, Marseille in the late thirteenth century was the largest city of the county of Provence and a commercial entrepôt of a stature nearly equal to that of Genoa.[5] Apart from the transshipment of merchandise, the practice responsible for the city's commercial fame, Marseille's economy revolved around ship-building, wine-production, livestock, fish, leather trades, and coral-work. Within the walls worked the usual spectrum of artisans, retailers, service trades, professional groups, office-holders, nobles, laborers, and fishermen, with a heavy concentration of leather-workers and members of seafaring professions. Agricultural workers tended to vineyards, fields, and gardens in the surrounding territory, shepherds pastured their flocks on the high hills surrounding the city, and carters and drovers kept up a steady flow of merchandise and animals between Marseille and the nearby cities of Aix and Avignon. The surrounding countryside was dotted with villages and fortified towns, the rural retreats of many of Marseille's military nobility; the highlands, in troubled times, sheltered villains who would kidnap travelers and send a messenger back to the city to demand ransom from

---

[2] Victor-L. Bourrilly, *Essai sur l'histoire politique de Marseille des origines à 1264* (Aix-en-Provence, 1925); Régine Pernoud, *Essai sur l'histoire du port de Marseille des origines à la fin du XIIIème siècle* (Marseille, 1935); Georges Lesage, *Marseille angevine: recherches sur son évolution administrative, économique et urbaine de la victoire de Charles d'Anjou à l'arrivée de Jeanne Ire (1264–1348)* (Paris,1950); Louis Blancard, *Documents inédits sur le commerce de Marseille au moyen âge, édités intégralement ou analysés*, 2 vols. (Geneva, 1978); John Pryor, *Business Contracts of Medieval Provence. Selected "Notulae" from the Cartulary of Giraud Amalric of Marseilles, 1248* (1884; Toronto, 1981).

[3] See below, n. 47.

[4] Chambre du commerce de Marseille, *Histoire du commerce de Marseille*, 7 vols, ed. Gaston Rambert (Paris, 1949–1966); Christian Maurel, "Structures familiales et solidarités lignagères à Marseille au XVe siècle: autour de l'ascension sociale des Forbin," *Annales ESC* 41 (1986): 657–81.

[5] For population estimates of medieval Marseille, see Édouard Baratier, *La démographie provençale du XIIIe au XVIe siècle* (Paris, 1961), 66, n. 1; *Histoire de Marseille*, 102; Daniel Lord Smail, "The General Taille of Marseille, 1360–1361: A Social and Demographic Study," *Provence historique* 49 (1999): 473–85.

their wives. Despite the absence of a genuine *contado* or territory, linkages between the city and the country were strong and communication regular. Marseille was the commercial hub of the region, so a constant flow of agricultural produce, particularly wine, made its way into the city for export.

Until the mid-fourteenth century, the city inside the walls was divided into three separate political jurisdictions, and several burgs constituted a fourth, politically unrecognized type of district. First in importance was the lower city (*ville inferioris*), stretching along the port and climbing part way up the northern hills. Above it, nestled in the apex of the triangle, lay the upper city (*ville superioris*). To the west lay the district known as the Prévôté, a small district surrounding the cathedral and under the rule of its canons. Administratively, the lower city was divided into six *sixains* or administrative sections similar in size to the *gonfalons* of Florence. Of these six, the only one with no access to the port was St. Martin. In council records they are always listed in geographical order from west to east: St. Jean, Accoules, Draparia, St. Jacques, St. Martin, and Callada. All took names from institutions located within their boundaries. Accoules and St. Martin were named after local parish churches, although the boundaries of parish and *sixain* did not correspond with each other. The Hospitalers of St. John of Jerusalem had their compound within the boundaries of St. Jean, and hence gave their name to the *sixain*, and St. Jacques bore the name of a church located within its boundaries, the church of St. Jacques de Corregaria. Draparia simply means "Drapery," or the Drapers' quarter, and the workshops of many drapers were indeed located in the *sixain*. Callada probably took its name from the word for paving stones (*calatae*).[6] In general, the *sixains* are encountered rarely outside the council records. Most people bore no powerful tie to their *sixain*. This lack of identification with administrative quarters, which is a general Provençal trait, stands in some contrast to the loyalties that could be generated in the quarters of Florence and other European towns.[7]

Starting from the west, the *sixain* of St. Jean was dominated by trades related to the sea, including fishermen, mariners, caulkers, canvasmakers, and oarmakers. In the *sixains* of Accoules and Draparia, merchants, nobles, bankers, and other notable figures such as judges were

---

[6] Charles Du Fresne Du Cange, *Glossarium mediae et infimae latinitatis* (Paris, 1842), s.v. "calatum."

[7] See, in particular, D. V. and F. W. Kent, *Neighbours and Neighbourhood in Renaissance Florence: The District of the Red Lion in the Fifteenth Century* (Locust Valley, N.Y., 1982); Nicholas A. Eckstein, *The District of the Green Dragon: Neighbourhood Life and Social Change in Renaissance Florence* (Florence, 1995); Jacques Heers, *Family Clans in the Middle Ages: A Study of Political and Social Structures in Urban Areas*, trans. Barry Herbert (Amsterdam, 1977), 129–68.

concentrated in a strip of land along the port that extended inland for two or three blocks. Further inland, up the hill, the *sixain* of Draparia was dominated by workers in cloth and leather trades, and important fish- and meat-markets were located within a few blocks of the plaza of Accoules, the heart of the city. Cobblers were found in important numbers in Callada, along with a great many agricultural laborers and, in the Carpentry, carpenters or shipwrights and painters.

The upper city, in turn, was divided administratively into four quarters: Rocabarbola, St. Jacques of the Swords, Porta Gallica, and St. Cannat. The council of the upper city, to judge by surviving records, lacked the independence and clout of its lower city counterpart, and the residents of the upper city paid even less attention to their administrative quarters than did their counterparts in the lower city. It is true that Rocabarbola is frequently encountered in the records and that people used the name to mean a geographical entity larger than the street of Rocabarbola. Since people did not use any of the names of the other quarters, however, it is unlikely that the use of "Rocabarbola," in popular usage, meant the administrative quarter. The other common area of the upper city in terms of usage is Cavalhon. The name does not refer to an administrative quarter, and the nature of the entity is not clear. The dozens of site clauses that use the name all refer to an area west of the street of the Upper Grain Market, east of the Prévôté, and north of St. Antoine.[8] The upper city in the mid-fourteenth century was dominated by agricultural laborers: 65 percent of the episcopal city and the Prévôté together were identified as laborers in various records, compared to 12 percent in the lower city.[9] Important exceptions include fishermen and a few nobles, and there was a sprinkling of artisans. The Prévôté, apart from laborers, fishermen, artisans, and merchants, was also home to an assortment of canons and other men attached to the cathedral.

The other important administrative division consisted of the numerous burgs or suburbs (*burgi, surburbia*) that ringed the city to the north and east. These first formed in the middle of the thirteenth century.[10] Starting from the north and moving clockwise, the order of the burgs most frequently encountered in the mid-fourteenth century is roughly as

[8] Many of the house in this area fell under the direct lordship of the bishop of Marseille, and hence frequent references to Cavalhon can be found in episcopal rent registers. See, for example, ADBR 5G 112, 5G 114, 5G 115, and 5G 116, from the 1340s through the 1360s.

[9] The source for this is the prosopographical index.

[10] Marc Bouiron, "Le fond du vieux-port à Marseille, des marécages à la place Général-de-Gaulle," *Méditerranée* 3 (1995): 65–68.

2. Marseille, ca. 1350.

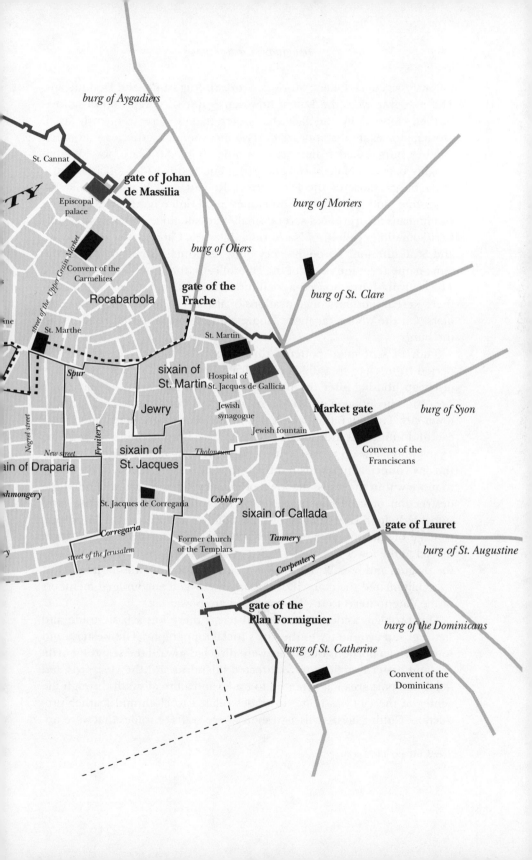

burg of Aygadiers

St. Cannat

gate of Johan de Massilia

burg of Moriers

Episcopal palace

burg of Oliers

Convent of the Carmelites

Rocabarbola

gate of the Frache

burg of St. Clare

St. Marthe

St. Martin

Spur

sixain of St. Martin

Hospital of St. Jacques de Gallicia

Market gate

burg of Syon

Jewry

Jewish synagogue

Convent of the Franciscans

Negrel street

Fruitery

New street

Jewish fountain

sixain of St. Jacques

Tholoneum

ain of Draparia

St. Jacques de Corregaria

Cobblery

sixain of Callada

gate of Lauret

shmongery

Corregaria

Former church of the Templars

Tannery

burg of St. Augustine

street of the Jerusalem

Carpentery

gate of the Plan Formiguier

burg of the Dominicans

burg of St. Catherine

Convent of the Dominicans

street of the Upper Grain Market

follows: Aygadiers, Oliers, Moriers, Syon, St. Augustine, and Dominicans. The topography of the burgs, however, is not very well known, and is further clouded by the fact that other names were frequently used. Moriers, for example, appears to have incorporated or at least adjoined another burg called Malemortis. Another burg, St. Clare, was situated roughly between Malemortis and Syon, and Syon shaded off into three other burgs, those of the Franciscans, Robaut, and Prat d'Auquier. St. Augustine and Syon were sometimes used interchangeably, and the Dominicans incorporated several smaller appellations, the burgs (as they were sometimes called) of Na Auriola, Pilas, Na Capona, Raymon Rascas, and St. Catherine. Through every burg ran a main street that bore the same name (e. g. the street of the Dominicans, the street of the Moriers). The outlying burgs housed agricultural laborers, and closer to the walls there were important concentrations of artisans such as tanners and cobblers. On the whole, far more laborers and fewer nobles were found in the burgs.

Study of settlement patterns throughout the urban area during the period from 1337 to 1362 is complicated by migration. The availability of cheap housing after the plague joined with the threat of war in the 1350s to encourage a great deal of migration out of burgs and into the lower city. This migration is revealed by the prosopographical index: one can find numerous individuals known to have lived in the suburbs in the 1340s who show up as residents of the lower city in the 1350s. In 1357 the city council actually invited thirty-four notable residents of the burgs to move within the walls: the logic of military defense recommended the destruction of suburban houses near the walls, and the invitation presumably helped overcome suburban political resistance.[11] A large percentage of the nobility and mercantile elite of the upper city, in turn, moved to the lower city after 1348, possibly for reasons having to do with the plague, and possibly because the formal union of cities in January of 1348 eliminated the council of the upper city and encouraged members of the elite to head to the lower city, where power lay.

This problem aside, in very broad terms there was a basic trade and status group asymmetry to the city's social topography. The western and southwestern portions of the city were directed toward the sea; the northern and eastern sections were directed inland toward the vineyards and fields; the port area was oriented to trade, and a broad swath through the center of the city was home to trades related to cloth and leather production. Cutting across this asymmetry were crafts or trades that were ori-

---

[11] AM BB 22, fols. 15r–v.

ented toward retailing or service, such as bakers, butchers, notaries, physicians, and barbers. These trades were spread with some uniformity throughout the city. The three or four markets of the city were distributed more or less evenly throughout the city on or near major thoroughfares, and the butchers and stalls of other provisioners were located in them; thus, isolated groups of butchers and provisioners clustered together despite the dispersion of the trade as a whole.

The geography of ecclesiastical establishments was similarly universal. To judge by testaments, the five parishes generated a certain amount of loyalty, for pious gifts to parish priests were common and people often chose to be buried in their parishes rather than in the more fashionable cemeteries of the religious orders. Three of the parishes bore the same names as *sixains*, although it is very unlikely that the boundaries of parish and *sixain* were identical. St. Laurent, the westernmost parish, served the population of the quarter of St. Jean, a clientele made up of fishermen and mariners; it is the only medieval parish church that is still substantially intact in modern Marseille. Notre Dame des Accoules was situated at the heart of the city and serviced the most notable clientele—the residents of the *sixains* of Accoules, Draparia, and St. Jacques. The church of St. Martin, in turn, was placed along the eastern wall, and included among its parishioners the residents of St. Martin and Callada and all the southwestern burgs. The upper city was served by the cathedral of La Major in the west and the parish church of St. Cannat in the north, near the apex of the triangle and the episcopal compound. The churches of St. Jacques de Corregaria, La Major, St. Catherine, the mendicant churches, and the churches associated with the hospitals also serviced the spiritual needs of a fair percentage of the population, and the cemeteries attached to some of these churches competed with those of the parochial churches for the bodies of the dead.

Apart from the venerable monastery of St. Victor, founded by St. John Cassian, other ecclesiastical establishments included the equally ancient female monastery of St. Sauveur, the cathedral chapter, the episcopal compound, the military religious order of the Hospitalers of St. John of Jerusalem, Carmelites, Franciscans, Dominicans, Augustinians, the brothers of St. Anthony, Clares, Beguines, and the hospital of St. Esprit. Before the suppression of the order, the Templars too had their church which was taken over by the Augustinians in the fourteenth century. In 1348 the new hospital of St. Jacques de Gallicia was founded by the merchant Bernat Garnier. There was a thriving Jewish quarter near St. Martin with a synagogue and an almshouse; some distance out in the countryside the Jews had their own cemetery.

These were the physical and political patterns inherited from Marseille's glorious past. "A city of princely and wise citizens," remarked a traveler, Benjamin of Tudela, in the 1170s, "it is a very busy city upon the seacoast."[12] Benjamin also noted the curious division of the city, at that time divided into an upper city that had been under the dominion of the bishop since the time of Gregory of Tours, and the lower city, under the dominion of the viscounts of Marseille. It was in the twelfth century that the commercially vibrant lower city first experienced the stirrings of communal independence.[13] By the year 1178, around the time of Benjamin's visit, we hear of a council of six men who assisted the viscounts in the administration of the lower city, and by the end of the century the move toward communal independence was well underway. In the year 1212, a confederation of leading members of the merchant guild formed the confraternity of the hospital of St. Esprit and began to buy off the viscountal rights, a process nearly complete by 1226, although the confraternity itself had disbanded in 1220.[14] As a result of this movement, the lower city achieved a measure of independence for two decades, in spite of the hostility of the bishop and lord of the upper city. In this early political order, craft guilds played a significant role, for the council included one hundred heads of trade groups elected by the members of each trade in addition to eighty-three councilors at large.[15] Jean Schneider has remarked that in French cities at a comparable moment in time, "family clans were never organized into a patriciate; instead, merchant guilds and craft groups often played a decisive role in the emancipation of cities."[16] Schneider's implicit contrast was with the cities of the Italian peninsula and the Low Countries where urban oligarchies formed around the leading families.[17] In this regard, thirteenth-century Marseille was more French than Italian. In 1221, the existing system was replaced by a foreign *podestat*, in the manner being pioneered by some Italian cities at the same moment; the system lasted until 1230.

---

[12] Benjamin of Tudela, *The Itinerary of Benjamin of Tudela: Travels in the Middle Ages*, ed. Michael A. Signer (Malibu, Calif., 1983), 62.

[13] For much of what follows, see Bourrilly, *Essai*, and Baratier, *Histoire de Marseille*.

[14] See Paul Amargier, "Mouvements populaires et confrérie du Saint-Esprit à Marseille au XIIIᶜ siècle," in *La religion populaire en Languedoc du XIIIᵉ siècle à la moitié du XIVᵉ*, vol. 11 of *Cahiers de Fanjeaux* (Toulouse, 1976), 305–19.

[15] Bourrilly, *Essai*, 206–10.

[16] Jean Schneider, "Problèmes d'histoire urbaine dans la France médiévale," in *Tendances, perspectives et méthodes de l'histoire médiévale*, vol. 1 of *Actes du 100ᵉ congrès national des sociétés savantes. Paris, 1975. Section de philologie et d'histoire jusqu'à 1610* (Paris, 1975): 147.

[17] See, among others, Jean Lestocquoy, *Aux origines de la bourgeoisie: les villes de Flandre et d'Italie sous le gouvernement des patriciens (XIᵉ–XVᵉ siècles)* (Paris, 1952).

In the wake of this political transformation, the viscounts of Marseille faded from view. Seigneurial claims did not, for the count of Provence and overlord of the viscounts, Raimond-Bérenger, continued to press his own claims on the lower city. In a strategic move to fend off pressure from Raimond-Bérenger, in 1230 the lower city of Marseille submitted itself to the count of Toulouse, Raimond VII, who had been politically weakened by the disastrous results of the Albigensian crusade. In return, the city received a certain measure of political autonomy. In 1246, however, a much more formidable enemy arose—Charles of Anjou, younger brother to Louis IX of France. Not content to remain in the shadow of his saintly brother, Charles would ultimately seize the kingdom of Naples and was eventually crowned king of Sicily and Jerusalem. In 1246, Charles married Beatrice, who to the exclusion of her sisters was sole heir of her father, Raimond-Bérenger. Her dowry included the county of Provence, of which the prize was Marseille itself. However, the cities of Provence did not take well to this and formed a league against Charles, though in 1250 the soon-to-be king arrived in Provence to press his claim. Military resistance ended by 1251, and in 1252 the lower city of Marseille signed a treaty acknowledging Charles' suzerainty. The terms of the treaty were not harsh. Marseille promised military aid to Charles and half its municipal revenues but retained a great deal of autonomy. Certain members of Marseille's mercantile elite resisted the changes wrought by Charles, however, and the years 1257, 1261, and 1263 were marked by uprisings. The last was led by the Manduel brothers, wealthy merchants of the city, in alliance with other leading figures. Whether these men were motivated by a sense of loyalty to the city's former independence, as the historians of Marseille would have one believe, or by complex reasons having little to do with Charles and a lot to do with internecine strife, which is more likely but unprovable, the effect was the same, for Charles finally lost patience and cut off their heads. In the meantime, in 1257, Charles had purchased the *ban* of the bishop of Marseille, thereby gaining jurisdiction over the upper city, and by 1264 the authority of the new count of Provence over both cities was no longer in question.

The treaty of 1252 absorbed intact the existing body of city statutes acquired during the period of communal independence, and Charles accorded the city numerous privileges. The succession of revolts had the effect of narrowing those privileges.[18] In particular, the treaty that

---

[18] Mireille Zarb, *Histoire d'une autonomie communale: les privilèges de la ville de Marseille du Xe siècle à la Révolution* (Paris, 1961).

followed the 1257 revolt, known as the Chapters of the Peace, imposed on the lower city a viguier (*vicarius*) and sub-viguier to be named by the king of Naples and took away from the municipal council the right to nominate judges. The guilds were henceforward excluded from all formal representation on the council—this was a backhanded reminder of their impressive degree of political power during the period of communal independence, as Georges Lesage remarks.[19] Last, the city lost the right to collect indirect revenue. Nevertheless, the city managed to retain significant privileges, notably the right to keep courts of first and second appeals within the city, and as a result the Chapters of the Peace, far from being dishonoring, came to be accepted as a major source of civic pride and identity. Subsequent redactions of the Chapters of the Peace made over the course of the fourteenth century included two beautiful miniatures, one depicting the presentation of the Chapters to the count of Provence, the second showing a viguier taking an oath with his hands on the book.[20] The upper city was left out of these treaties and remained under the direct rule of the count until Queen Jeanne unified the upper and lower cities in January of 1348.

A significant series of registers of the deliberations of Marseille's city council has survived from 1318 to the end of the century, and from these we can get a sense of how the political system actually came to function in Marseille.[21] According to the Chapters of the Peace, the viguier, in principle, was the highest authority in the city, and a direct representative of the crown. The viguier was invariably a foreigner to Marseille, typically a knight and lord from elsewhere in the county of Provence, brought in for the space of a year to serve as the city's nominal leader. The viguier governed with the aid of prominent citizens of the city, and to this end nominated six prud'hommes. These men were invariably drawn from the members of the council, and their offices, like that of the viguier, were rotated on a yearly basis. Among their other duties, the six were in turn responsible for nominating the members of the council from among the leading citizens of each of the *sixains* of the lower city. They also supervised the elections of the three city syndics, officials who increasingly took on duties nominally in the hands of the viguier, and chose the men who served on the many committees that oversaw the guilds and other civic concerns, such as prostitution and sanitation. Despite appearances, the

---

[19] Lesage, *Marseille angevine*, 40.
[20] AM AA 1 and AM AA 2.
[21] AM, series BB. These have been very carefully inventoried; see *Inventaire*. For the political system of fourteenth-century Marseille, see also Lesage, *Marseille angevine*, 66–69.

sixains were not given an equal voice on the council. Over a thirty year period from 1331 to 1361, the *sixain* of Draparia, certainly the most powerful of the six and possibly the most populous, averaged thirty-five or so council members each year, whereas the weakest and poorest *sixain*, St. Jean, managed an average of only nineteen. Despite the manifest imbalance, efforts were made to allow all *sixains* an equal voice on some matters, for committee assignments were made on the basis of one man, sometimes two, from each *sixain*. The offices of the six prud'hommes were also assigned to each of the six *sixains*. Again, this appearance of fairness can be deceptive, for the two men who represented St. Martin and Callada in the year 1330 turn out, in 1331, to be councilors representing the *sixains* of Draparia and St. Jacques, giving every appearance that Draparia and St. Jacques stacked things in their favor in the previous year. This was possible because ownership of property and not actual residence seems to have determined where one could choose to place one's *sixain* loyalties.

Despite the formal exclusion of the guilds from governance, the council was not entirely restricted to nobles and great merchants, for it did include a fair percentage of artisans and professionals—men of wealth and standing, to be sure, but common people, nonetheless. The council register for the year 1350–51, which lists 199 members of the general council, includes, among the 165 whose professions can be readily identified from other sources found in the prosopographical index, four apothecaries, three butchers, two carpenters, one cobbler, one fisherman, nine laborers, two mariners, three notaries, one oarmaker, one shoemaker, and one ship-steward.[22] Close to 17 percent of the total number of councilors, therefore, were men who had their origins in crafts or trades. Furthermore, a great number of the 145 positions on the forty-eight city committees, ranging from the committee on the fair measuring of wine (*Mensuratores vini*) to the committee on expelling prostitutes from good neighborhoods (*Ad expellandum meretrices*) were regularly filled by men from a wide range of professions. Before the Black Death, things were not so democratic. In 1339, among the sixty-seven councilors of known profession or status, there was one apothecary, one cooper, one cobbler, one carpenter, one mariner, and one hatter; tradesmen, in other words, amounted to only 9 percent of the council body in 1339.[23] Even

---

[22] AM BB 21, fols. 1–18. I exclude merchants, drapers, and moneychangers from this list, as these were decidedly prestigious professions and included many members who styled themselves as nobles.

[23] AM BB 19, fols. 1r–14v.

so, although the council was decidedly oligarchic, this was hardly like the situation in Venice where the statute of 1297 and its successors tried, with some success, to limit future participation in the council to the descendants of those nobles and recent immigrants of noble status who were sitting or invited to join in 1297.[24]

In Marseille, the judicial system was led by an Angevin-appointed official known as the palace judge, head of a civil court of first instance called the palace court.[25] One of the most significant rights conceded by Charles to the city was the privilege of *non extrahendo*, the right to keep all judicial appeals within the city walls, and so the palace court was backed up by two courts of first and second appeals. All three judges, like the viguier, were foreigners, nominees of the distant Angevin crown in Naples, and, like all city offices, the positions were rotated on a yearly basis. At some point before the mid-fourteenth century, two lesser civil courts staffed by local jurists were added to the roster. These were given simple, no-nonsense names: "Marseille's court" (*curia Massilie*) and "Marseille's other court" (*curia alter Massilie*). The upper city, until 1348, had a court similar in nature to the palace court headed by the *judex Turrium* or "the judge of the city of Towers," and the episcopal court, also located in the upper city, was active. Numerous records of the palace court, the court of first appeals, and the two lesser courts have survived from the mid-fourteenth century. These documents are among the best judicial records to survive from fourteenth-century France and include long sections of witness testimony that sometimes show how people thought about and imagined their world.

By the mid-fourteenth century there was also a very active criminal court of inquest (*curia inquisitionis*), its origins dating to the mid thirteenth. The court of inquest was headed by the palace judge, assisted by the two judges of the lesser courts; it had its own notary and kept its registers distinct from those of the palace court. No registers of the court of inquest are known to be extant before 1380, so we cannot examine the procedures as closely as we might like. There are two sources, however, which give a sense of its caseload. First, a list of fines paid for criminal

---

[24] Frederick C. Lane, "The Enlargement of the Great Council of Venice," in *Florilegium Historiale: Essays Presented to Wallace K. Ferguson*, ed. J. G. Rowe and W. H. Stockdale (Toronto, 1971), 237–74.

[25] A serviceable introduction to the courts of Marseille can be found in Raoul Busquet, "L'organisation de la justice à Marseille au moyen âge," *Provincia* 2 (1922): 1–15. See also my "Notaries, Courts, and the Legal Culture of Late Medieval Marseille," in *Urban and Rural Communities in Medieval France: Provence and Languedoc, 1000–1500*, ed. Kathryn L. Reyerson and John Drendel (Leiden, 1998), 23–50.

condemnations has survived from the fiscal year 1330–1331. To judge by this list, the courts of inquest for both the upper and lower city were together processing close to 500 cases per year, chiefly small matters involving fights, insults, and the bearing of illegal weapons.[26] Second, appeals heard before the court of first appeals often included transcripts of the original inquest.

The two decades that followed the last revolt of 1263 were decades of relative peace and prosperity. Building on its monopoly over the transshipment of goods coming down the Rhone, Marseille continued to exploit a trading network that its merchants had established across the Mediterranean, chiefly in the Levant and the north coast of Africa. And then, in 1282, disaster struck with the Vespers, the Sicilian revolt against Angevin rule. The loss of Sicily bit deep into the core of Angevin power in the western Mediterranean and precipitated a long period of Angevin decay. By 1282, the political web that tied Marseille to Naples, the seat of the Angevin throne, was sufficiently strong to drag down Marseille as well.[27] According to Édouard Baratier, Marseille's shipyards were diverted to the production of warships to aid the Angevin cause, and the commercial fleet suffered as a result.[28] The economic contraction from the late thirteenth century onward, it is true, was a general Mediterranean phenomenon: the Sicilian Vespers was a symptom of this decline, not a cause. The changing focus of the fairs of Champagne after 1260, coupled with the fall of Acre in 1291, restricted the range of Mediterranean markets available to the merchants of Marseille, and Montpellier began to emerge as the chief port city along the French and Provençal littoral.[29] By 1348, Marseille's trade was limited to the corner of the western Mediterranean nestled between the Iberian peninsula to the west and the Italian peninsula to the east, with the chief axes radiating out to Genoa, Naples, Sardinia, the French and Provençal coastal towns, and Barcelona, although the increasingly tense relations with Catalonia that culminated in the sack of 1423 eventually eliminated the western axis.

---

[26] ADBR B 1940, fols. 74r–139v.

[27] See *Marseille et ses rois de Naples: la diagonale angevine, 1265–1382* (Aix-en-Provence, 1988).

[28] Baratier, *Histoire de Marseille*, 95–96.

[29] See, in general, Édouard Baratier and Félix Reynaud, *De 1291 à 1480*, vol. 2 of *Histoire du commerce de Marseille* (Paris, 1951); M. M. Postan and Edward Miller, ed., *The Cambridge Economic History*, 2d ed., 3 vols. (Cambridge, 1987), 2:341–43. On the changing focus of the fairs of Champagne, see Robert-Henri Bautier, "Les foires de Champagne: recherches sur une évolution historique," in *La foire*, vol. 5 of *Recueils de la Société Jean Bodin* (Brussels, 1953), 97–147.

Piracy cut deeply into profits. Bandits lurked in the hinterland, preying on caravans, capturing men and holding them for ransom. The second half of the fourteenth century was particularly troubled, especially after Queen Jeanne succeeded her father, Robert, to the throne in 1342. Intrigues at the court encouraged political instability throughout Provence, pitting Marseille, loyal to Queen Jeanne, against Aix, Arles, and other Provençal cities.[30] In the late 1350s, northern politics spilled over into the south. The city was threatened by roving bands of mercenaries, cast adrift from the French army following the spectacular English victory at the Battle of Poitiers in 1356, a major episode in the Hundred Years War. The year 1357 was especially troubled by the ravages of a French war leader turned brigand known as the Archpriest, who, with his men, turned to plunder to recoup his losses in battle. The council record for that year is filled with evidence of the defensive measures taken.[31]

In the midst of these troubles, the commercial decline was precipitous. In 1248, a single notarial casebook contained more than a thousand acts of a commercial nature.[32] In *all* the surviving notarial records for the entire decade preceding 1348—some thirty casebooks numbering thousands of folios—there are only 147 commercial acts. So great was the despair that in 1345 the Angevin crown sponsored an inquiry into the reasons for diminishing royal revenues in the once wealthy port of Marseille. The testimony of a series of great merchants and armateurs pointed to the same things: a loss of Levantine markets and the effects of piracy.[33]

But if Marseille's economy collapsed with the declining Angevin presence in the Mediterranean that followed upon the Sicilian Vespers, the city's political life followed a different trajectory, for the weakened condition of the Angevin crown allowed Marseille's council quietly to recover a measure of independence. To judge by the council records from the fourteenth century, the Angevin viguier had become a cipher, deferring to the council and its representative bodies in all but a few matters that concerned the Angevins directly. The real source of power in the mid-fourteenth-century city lay with the council and in particular with the

---

[30] Baratier, *Histoire de Marseille*, 103–4; Émile-G. Léonard, *Histoire de Jeanne Ière, reine de Naples, comtesse de Provence (1343–1382)*, 3 vols. (Monaco, 1932).

[31] *Inventaire*, 68–82.

[32] John Pryor, *Business Contracts of Medieval Provence: Selected "Notulae" from the Cartulary of Giraud Amalric of Marseilles, 1248* (Toronto, 1981), 41. Rosalind Kent Berlow gives a sense of the sheer scale of trade in her "The Sailing of the 'Saint Esprit,' " *Journal of Economic History* 39 (1979): 345–62, which is based on the registers of Giraud.

[33] Transcribed in Georges Lesage, *Marseille angevine*, 184–86.

syndics. In this changing political environment, an oligarchy of merchants and nobles was in the process of shaping itself in the middle decades of the fourteenth century and quietly recapturing a measure of the city's former independence.[34] Many members of the nobility dealt in commerce or armaments, and wealthy merchants frequently invested their profits in rents. They married one another and often lived near one another. On paper, this oligarchy had a profound and touching loyalty to the Angevin crown and the cult of the Angevin saint, St. Louis d'Anjou. It consistently represented itself in the pages of the council deliberations through an ideology of civic harmony. These were polite fictions, for court records reveal a feud of major proportions that split the oligarchy into groups of bitter enemies. The rhetoric that surfaced during attempts to prosecute the feud denied implicitly the reality of Angevin hegemony.[35]

However tacit, the two factions were powerful political structures. They conform to no clear socio-economic pattern, for both groups included members of the military nobility and mercantile elite as well as an indiscriminate assortment of artisans and laborers. This is not to say that there was no pattern behind it all. Among other things, it is quite clear that neighborhood rivalries between powerful individuals or families were absorbed into the larger factional hatred; put differently, factional allegiance was one way in which neighborhood rivals, like the Toesco and the Engles, or the Mercier and the de Serviers, competed for status. One result of this is that if one pinpoints all the domiciles of the members of one party on a map, the resulting image is almost indistinguishable from a similar map drawn up for the other party.[36] Second, those few men on the city council who did *not* yield to the tempting sport of vengeance and warfare were those whose world of kin was among the smallest. The great Peire Austria, probably the most successful merchant of his day, was the son of an immigrant. He was childless, and his two younger brothers Salvayron and Ludovicet—always mentioned in tandem, always with the diminutive names—were so inactive as to appear half-witted. Johan de Sant Jacme, a leading city councilor and a peacemaker besides, may have been, along with his brother Raynaut, a recent transplant from the countryside; neither had any known relatives in the city. The great merchant Bernat Garnier was childless; upon his death in 1347, his fantastic wealth

---

[34] See in general Maurel, "Structures familiales"; idem, "Le prince et la cité: Marseille et ses rois . . . de Naples (fin XIIIe–fin XIVe siècles)," in *Marseille et ses rois de Naples*, 91–98.
[35] See my "Telling Tales in Angevin Courts," *French Historical Studies* 20 (1997): 183–215.
[36] See my "Mapping Networks and Knowledge in Medieval Marseille: Variations on a Theme of Mobility" (Ph.D. diss., University of Michigan, 1994), 145–62.

went to found the Hospital of St. Jacques de Gallicia. Exposed and kinless, these men were powerless, and hence they sought to make their honor elsewhere.

At the other end of the social and political spectrum of mid-fourteenth-century Marseille was a world of impecunious laborers, ruffians, thieves, murderers, prostitutes, lepers, and abandoned children. We learn of them from time to time, but only in passing, or sometimes in rumor. If they swelled the population of Marseille, they did not fill the pages of the existing sources, because the regulation of their world in the mid-fourteenth century was only just emerging as a major concern of the courts and the council.[37] In between lay the world of respectable laborers, fishermen, and mariners; tradesmen and tradeswomen, retailers, and artisans; physicians, notaries, jurists, and schoolmasters; bankers and moneylenders. Their formal access to power was limited, although the wealthiest or most prominent men among them could rise to serve on the council or on one of the committees appended to the council. They, too, fought and feuded among themselves, even if the courts made efforts to take the dignity out of these battles. They married across neighborhoods and professional groupings; they moved about the city and countryside with frequency and ease; they constructed multitudinous networks of knowledge among a variety of acquaintances. Regardless of their origin, they dealt freely with the notaries and the law, using it for a wide variety of purposes. They borrowed and traded, and when they died they paid for clergymen to carry their bodies and sing masses for their souls.

THE CITY IMAGINED

If the city was a real place, a subject of social and political history, so too was it a site of imagination. I turn now briefly to the city imagined, and begin with its own sense of its history. Like any other city, Marseille did not have a single past, and we can sense the conflicts of the thirteenth century surfacing in the politics of the fourteenth. To judge by Benjamin of Tudela's remarks from the late twelfth century, where he speaks of the thriving Jewish academies and the distinguished Jewish merchants who

---

[37] A reading of the deliberations of the city council reveals that an interest in regulating the world of marginals was developing across the fourteenth century; see *Inventaire*. See also Bronislaw Geremek, *The Margins of Society in Late Medieval Paris*, trans. Jean Birrell (Cambridge, 1987); R. I. Moore, *Formation of a Persecuting Society: Power and Deviance in Western Europe, 950–1250* (New York, 1987); Jacques Rossiaud, *Medieval Prostitution*, trans. Lydia G. Cochrane (New York, 1988).

helped make the commercial revolution, and to judge in turn by the potsherds of the memories of the Jewish past that survive in the documents of fourteenth-century Marseille, the Jews had their own history. So did craft groups, to go by the profoundly craft-oriented geography that continued to shape the map of fourteenth-century Marseille. The great merchant inquest of 1345 tells us of the memories of a merchant past, a past made memorable by the immense profits that were to be had, a period of merchant fortunes that had shuddered to a halt—they agreed on this point—with the Muslim reconquest of Acre in 1291 and the more limited trading horizons that ensued. Memories dredged up during the trials of members of the great feuding families of Marseille reveal the historical depth of hatred. This is not to speak of those monks and friars whose histories are almost entirely inaccessible to us in the documents at hand, nor of the patricians of the fourteenth century who were chary about discussing any memories of thirteenth-century resistance which they may have shared among themselves in their painted urban parlors and rural retreats.

If these histories are mostly lost to us, they leave a legacy in the toponyms of the fourteenth century. In some fundamental sense the cartographic imagination is nothing more than one of many ways in which people convey an image of the past in the confines of the present. To take one example, certain vicinities took names from artisanal groups or retail centers, such as the Smithery, the Carpentery, or the Tripery. The usage of the names in the mid-fourteenth century, however, bears only a partial relationship to anything one might call the real world, since some names of this type continued to be used after the artisans or retailers in question had drifted elsewhere. Other areas in which members of a trade were concentrated did not always acquire the name of the craft. Similarly, there is a play of power, for the use (or non-use) of a craft-based name may indicate a political contest over the importance of craft identity to the civic body.

So we find that toponymy often departs from reality and does not reflect the ever-changing situation of the city. We see this notably in the expressions used to described the city itself, or rather the two distinct political entities, the upper and lower cities, that were the chief elements of Marseille before the union of 1348. The names of the two cities were not fixed. Nostalgia often prompted scribes to call the lower city the "city of the Viscounts" (*ville vicecomitalis*), a century and more after the last of the viscountal rights had been bought off; greatly tempted to borrow the more euphonius name, I have used the more prosaic form ("the lower city") only because it is somewhat more common. The upper city, the *ville superioris,* also had its cognomens, for the scribes continued to call it "the Episcopal city" (*ville episcopalis*) a century after the bishop had sold his rights to Charles of Anjou. Occasionally we even read of the "city of Towers" (*ville Turrium*), a name linking the upper city to a distant past when corporate agnatic lineages built fortifications and fought among themselves. The romance fades quickly upon closer inspection, for some of the towers were in ruins, and few were conspicuous elements of patrimonies attached to patrilineages. Some were owned by laborers, fishermen, and butchers. In 1355, one was being sold by two women.[38]

The continuing use of such names long past their historical relevance suggests that toponymy could be conservative and nostalgic. There are other examples. The terms *ville inferioris* and *ville superioris,* for example, outlasted the formal unification of the cities in January 1348. The history of difference was kept alive by the obvious hostilities that could and did separate the upper and lower cities.[39] Certain episcopal officials continued to speak of the Jewry of the upper city at a time when all other evidence suggests that the Jews of Marseille had congregated in the Jewry of the lower city.[40]

Some of the most enduring cartographic features of the city, for this reason, were the very buildings and monuments that figure in the landmarks so favored by vernacular linguistic cartography. Stone was the stuff from which the Mediterranean city was made; the city itself began at its walls. A wall was civilizing, an imposing mass that separated the city from the countryside.[41] But walls are made to be penetrated, and six gates broke the city walls. The Gallican gate, the gate of Johan de Massilia, the gate of the Frache, the gate of Tholoneum or the Market gate, the gate of Lauret or of the Dominicans, and the gate of the Plan Formiguier— these were names of long standing. They loomed large in the minds of residents and city councilors alike, and it is significant that the wall, such

---

[38] ADBR 351E 4, fols. 27r–28r, 29 Sept. 1355.

[39] The minutes of council meetings record numerous disputes between the two cities. For several examples from the year 1332 concerning disputes over infrastructural repairs, passageways, wheat imports, and the ban on the exportation of building materials, see *Inventaire,* 35–36.

[40] Daniel Lord Smail, "The Two Synagogues of Medieval Marseille: Documentary Evidence," *Revue des études juives* 154 (1995): 115–24.

[41] Jacques Heers, ed., *Fortifications, portes de villes, places publiques, dans le monde méditerranéen* (Paris, 1985); see also idem, *La ville au moyen âge en Occident* (Paris, 1990), 328–31. On the symbolic meaning of walls in artistic representations, see Chiara Frugoni, *A Distant City: Images of Urban Experience in the Medieval World,* trans. William McCuaig (Princeton, 1991), 11.

a defining element of city geography, was associated so profoundly with those points where it wasn't a wall. The gates were places that influenced the cartographic practices of the many people who lived or transacted business within a few blocks of each. These six were not the only gates. Within the city was a little gate, called the gate of the Jewry (*portale Jusatarie*), sometimes a gatelet (*portaletus Jusatarie*); it was the formal entrance to the Jewry of the lower city.[42] The Jewry, as it happens, could be penetrated from many directions; the gatelet merely marked its symbolic opening to the Christian world. A single document refers to a counterpart, similarly called the *portale Judaycum*, once attached to the moribund Jewry of the upper city near the bishop's palace.[43]

So it is easy for us to share the concern of the city council whenever it turned to look at the walls of the city in their disrepair, to order measures to be taken to fix the damage, to contemplate, even momentarily, the horror of being breached. Even more can we appreciate the situation of a port city, a commercial city, a city that by definition was open to the sea, porous at its very heart. Hence the great iron chain, slung across the mouth of the port like some gigantic chastity belt, protecting the city from unwanted penetration. No one knows when it was first forged—probably it was made in the thirteenth century. In the council records that have survived from the fourteenth century it appears from time to time, but always in disrepair.[44] Amid the welter of issues addressed by the council over the course of the fourteenth century, this concern for the well-being of the chain stands out rather oddly. In the minds of the city fathers, the greatest disaster to befall Marseille was not the Black Death of 1348. That only took the lives of people, and in any event was a memory that had to be shared with the rest of Europe. Rather, the greatest catastrophe was the sack of 1423, when marauding Catalans ravished and burnt the dispirited city. They took away the great iron chain which still hangs today in the cathedral of Valencia.[45] Its loss, it seems, is still a source of shame.

---

[42] Smail, "The Two Synagogues," 18–19, note 15; ADBR 355E 293, fols. 103r–105v, 13 June 1362.

[43] ADBR 351E 5, fol. 1r, Jan. 1361.

[44] E.g., *Inventaire*, 66.

[45] Busquet, *Histoire de Marseille*, 141–45; Baratier, *Histoire de Marseille*, 115. Baratier made a trip to Valencia to photograph the chain; the plate appears in Baratier and Reynaud, *De 1291 à 1480*, 368. The Marseille industrialist and *érudit* Bruno Roberty intended a study of the great sack of Marseille to be the fruit of his years of work on the topography and society of medieval Marseille, a project that was never completed. See the "Fonds Roberty," housed in ADBR 22F.

If the medieval city was a site of imagination in its own day, so too has it been the subject for the historical imagination. Like all European cities, Marseille has been the subject of fond historical remembrance that generated numerous local histories of the city, starting with the great *Histoire de Marseille* by Antoine Ruffi in the seventeenth century.[46] It is a curious and largely unremarked fact that the historical remembrance of many nineteenth- and early twentieth-century local urban historians in Europe often concerned the recovery of the street plans of their cities. Historical Marseille is no exception: three major published works and two unpublished projects have over the course of the nineteenth and the first half of the twentieth centuries, focused either partially or primarily on the streets of premodern Marseille.[47] The impulse has not abated. Even now, if you wander through the departmental archives in Marseille and peer over people's shoulders, you'll find numerous amateur historians and genealogists working with maps. A more systematic study would reveal the tight relationship that exists between genealogical research and topographic curiosity.

The first of Marseille's many topographic historians was Augustin Fabre, who published his *Notice historique sur les anciennes rues de Marseille, démolies en 1862 pour la création de la rue Impériale* in 1862 and included a few acerbic comments about the value of progress and civilization, however necessary these might be. Pleased, as he put it, by the book's favorable reception, he then undertook a slightly larger project, entitled simply *Les rues de Marseille*, the five volumes of which appeared five years later and included a great deal of material not only on the medieval city but on everything up to the present as well. Work undertaken by J. A. B. Mortreuil, Octave Teissier, and Philippe Mabilly over the next few decades on the topography of medieval Marseille shows how important the project was to Marseille's *érudits* in the closing years of the nineteenth century and into the twentieth. The project was subsequently recapitulated by Bruno Roberty (d. 1950), a Marseille industrialist who later in life became

---

[46] Antoine Ruffi, *Histoire de la ville de Marseille* (Marseille, 1652).

[47] Augustin Fabre, *Notice historique sur les anciennes rues de Marseille, démolies en 1862 pour la création de la rue Impériale* (Marseille, 1862); idem, *Les rues de Marseille*, 5 vols. (Marseille, 1867–69); Octave Teissier, *Marseille au moyen âge. Institutions municipales, topographies, plan de restitution de la ville* (Marseille, 1891); Philippe Mabilly, *Les villes de Marseille au moyen âge. Ville supérieure et ville de la prévôté* (Marseille, 1905). J. A. B. Mortreuil produced an onionskin map of the medieval city at some point in the late nineteenth century; see ADBR Fi 50. See also the Fonds Roberty, ADBR 22F, for the topographical work of Bruno Roberty. In more recent times, see Adrien Blès, *Dictionnaire historique des rues de Marseille* (Marseille, 1989).

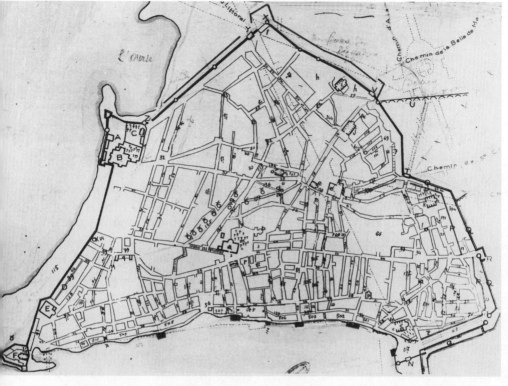

3. The Roberty map of Marseille, ca. 1423, detail.

a genealogist, archeologist, historian, and curator of the museum of archeology in Marseille, and it was his map of the city as it stood in 1423 that became a standard in histories of the city (Figure 3). The map was based on monumental labor; the papers he left to the departmental archives take up about twenty pages in the archival inventory and one item alone, his index to the streets, is about a foot thick.[48]

How did Roberty and his predecessors go about recovering the map of premodern Marseille? The basic street plan of the city probably remained essentially unchanged between the thirteenth and the sixteenth centuries and, despite the expansion of the city in the seventeenth century, many of the old streets remained in place. This, at any rate, was the experience of many medieval European cities, and there is no obvious reason to think

[48] ADBR 22F 86.

Marseille differed from the norm in any important respect. The street plan we can draw of medieval Marseille is based on early modern cadastral-type maps, but it is possible to hypothesize a principle of continuity and have it make sense in a medieval context. Starting with the known plan of early modern Marseille, these cartographer-historians were able, using notarial and other sources, to trace the names of streets back through the generations and find that medieval topographical descriptions worked reasonably well when laid on top of early modern maps.[49] This project, of course, reveals nothing so much as the fluidity of street names across the centuries, itself an interesting comment on the cartographic imagination of premodern Europe.

A striking feature of all the maps produced by Fabre and his successors is that they focused on streets. To think only in terms of streets is to look for streets in the records and, of course, to find them, or at any rate, to find things that look like streets. But the documents do not always speak about streets, and they do not always use the familiar nesting of geopolitical entities that is found in the modern address. Instead, the sources will, from time to time, talk about islands, or vicinities, or burgs. These varied usages indicate that basic cartographic templates used by men and women in medieval Marseille did not always identify addresses or plats by means of streets, and people rarely developed addresses based on a nesting set of geopolitical entities. In the three chapters that follow, then, we will see how it came about that streets became the favored object of urban cartographic imagining, and explore the various other templates that, at one time, competed with streets as the basic units of cartographic awareness.

---

[49] Derek Keene discusses the method of topographical reconstruction in *Survey of Medieval Winchester*, 2 vols. (Oxford, 1985), 1:37–40.

CHAPTER TWO

The Notary as Cartographer

I n the work of the economic, religious, social, and legal historians of medieval and Renaissance Europe who use notarial casebooks, it is common to find the sources carved up into their constituent acts.[1] Here we find commercial contracts of various types; there, testaments; elsewhere, dotal acts, estate inventories, estate divisions, peace acts, arbitrations, and so on.[2] The strategy is a reasonable one for dealing

[1] Useful introductions to using notarial sources for social and economic history include Robert Latouche, "Étude sur le notariat dans le comté de Nice pendant le moyen âge," *Le moyen âge* 37 (1927): 129–69; R. Aubenas, *Étude sur le notariat provençal au moyen âge et sous l'Ancien Régime* (Aix-en-Provence, 1931); Diane Owen Hughes, "Toward Historical Ethnography: Notarial Records and Family History in the Middle Ages," *Historical Methods Newsletter* 7 (1974): 61–71; Bernard Vogler, ed., *Les actes notariés, source de l'histoire sociale, XVIe–XIXe*, actes du colloque de Strasbourg, mai *1978* (Strasbourg, 1979); Paolo Brezzi and Egmont Lee, eds., *Sources of Social History: Private Acts of the Late Middle Ages* (Toronto, 1980); John Pryor, *Business Contracts of Medieval Provence. Selected "Notulae" from the Cartulary of Giraud Amalric of Marseilles, 1248* (Toronto, 1981); John Drendel, "Notarial Practice in Rural Provence in the Early Fourteenth Century," in *Urban and Rural Communities in Medieval France: Provence and Languedoc, 1000–1500*, ed. Kathryn L. Reyerson and John Drendel (Leiden, 1998), 209–35. Notarial archives in southern France are described and listed in R. H. Bautier and Janine Sornay, *Les sources de l'histoire économique et sociale du moyen âge*, vol. 2 (Paris, 1968–84).
[2] A representative sample includes many of the works of David Herlihy, such as *Pisa in the Early Renaissance: A Study of Urban Growth* (New Haven, 1958) and *Medieval and Renaissance Pistoia: The Social History of an Italian Town, 1200–1430* (New Haven, 1967); P. L. Malaussena, *La vie en Provence orientale aux XIVe et XVe siècles. Un exemple: Grasse à travers les actes notariés* (Paris, 1969); Robert S. Lopez, *The Commercial Revolution of the Middle Ages, 950–1350* (Englewood Cliffs, N.J., 1971); Jacques Chiffoleau, *La comptabilité de l'au-delà: les hommes, la mort et la religion dans la région d'Avignon au quatorzième siècle* (Rome, 1980); Thomas J. Kuehn, *Emancipation in Late Medieval Florence* (New Brunswick, N.J., 1982) and *Law, Family, and Women: Toward a Legal Anthropology of Renaissance Italy* (Chicago, 1991); Kathryn L. Reyerson, *Business, Banking and Finance in Medieval Montpellier* (Toronto, 1985).

with the superabundance of notarial archives.[3] The acts analyzed in this
literature have an obvious interest, and account for a certain percentage
of notarial business: 15 percent in Marseille, to be precise. Other acts,
such as property transactions, sales of various types, and especially the
debt-related contracts that constitute by far the bulk of notarial labors,
have drawn less attention.[4]

Yet to atomize notarial casebooks in such a way ignores the social, legal,
and epistemological frameworks that informed the acts the notary produced and thereby eliminates the agency of the notary.[5] It is also to over-
look the little details peripheral to the legal thrust of any given set of acts.
Residing in the margins of meaning, often as much in silence as in presence, these details come to our attention accidentally, incrementally, and
sequentially. They emerge in the ways in which women and men chose,
or did not choose, to identify themselves, in the power structures that
underlay how clients and notaries selected witnesses, in the spaces used
for transacting business, in styles of dating and ideas about time, in the
evidence offered for the mobility of people and of the law, in the technologies of communication, memory, and knowledge conveyed by the
acts, in the culture of liquidity and debt that they reveal so well. Details
such as these speak to us about matters that have little to do with commerce, piety, family structure, or legal behavior. They tell us, however
brokenly, about memories and cosmologies, ideology and power. Only
through a whole-casebook approach to reading notarial archives can we
hope to address these themes, to reinsert the notary into the culture that
he helped produce.

[3] See the discussion in Brezzi and Lee, *Sources*, xxi–xxii.
[4] Exceptions include the following. Property conveyances are discussed in Kathryn L.
Reyerson, "Land, Houses and Real Estate Investment in Montpellier: A Study of the Notarial Property Transactions, 1293–1348," *Studies in Medieval and Renaissance History* 6 (1983):
39–112. Debts and related issues are discussed in many works, including Richard W. Emery,
The Jews of Perpignan in the Thirteenth Century: An Economic Study Based on Notarial Records (New
York, 1959); Malaussena, *Vie*, 254–63; Monique Wernham, *La communauté juive de Salon-de-
Provence d'après les actes notariés* (Toronto, 1987), 107–80; Noël Coulet, *Aix-en-Provence: éspace
et relations d'une capitale (milieu XIVᵉ siècle-milieu XVᵉ siècle)*, 2 vols. (Aix-en-Provence, 1988),
2:501–36; William Chester Jordan, *Women and Credit in Pre-industrial and Developing Societies*
(Philadelphia, 1993); Kathryn L. Reyerson, *Society, Law, and Trade in Medieval Montpellier*
(Aldershot, 1995).
[5] Historians of the notaries are insistent on this point. Of those I consulted, among the most
helpful were Jean-Paul Poisson, *Notaires et société: travaux d'histoire et de sociologie notariales*,
2 vols. (Paris, 1985–1990) and *Études notariales* (Paris, 1996); Philippe Godding, ed., *Le
notariat en roman pays de Brabant et l'enseignement du notariat à l'université catholique de Louvain*
(Brussels, 1980); Jean L. Laffont, ed., *Notaires, notariat et société sous l'ancien régime. Actes du
colloque de Toulouse, 15 et 16 décembre 1989* (Toulouse, 1990). Laffont is also the general editor
of an important series entitled "Histoire notariale," published by the Presses Universitaires
du Mirail in Toulouse.

Fragments of the imaginary maps of medieval Marseille also reside in these, the margins of meaning. We find them in descriptions of property sites, in the linguistic devices used to identify streets, alleys, squares, vicinities, and landmarks, in the residential component of identity clauses. To gather these indices in serial fashion and to rebuild the cartography that informed their writing is to study the notary not just as a witness or a scribe, but also as a cartographer, one who acted upon and shaped the cartographic imagination of his clients. To say that notaries were cartographers is not to say that they were conscious of creating and applying an expert science of cartography to their mental map of the city. From this distance in time it is impossible to guess what they were thinking, nor have I ever seen a document suggesting that Marseille's notaries ever reflected on the cartography that informed the clauses they drew up. But in another sense they were cartographers, because, more than anyone else in medieval Marseille, they were mastering techniques for identifying the locations of people and property in the city. These were verbal techniques, to be sure, but for all that they were abstract representations of space. What is more, notarial apprehension of space was—or was becoming—systematic and regular, and arguably was influencing and transforming the ways in which other people perceived urban space.

Cartographic information resides in three types of clauses found in notarial acts. The first is the personal identity clause. Located near the head of any notarial act, the personal identity clause names the parties involved and often provides identifying details including, from time to time, an "address" or place of residence. Second is the subscription clause; located at the end of the act, the subscription clause includes the name and sometimes the seal of the notary, the names of the witnesses, and the place where the act was made. Personal identity clauses and subscription clauses are found in all notarial acts. A third type of clause, the property site clause, is found only in acts involving the conveyance of rights in real property; this clause identifies the location of the property. Here, I will be concerned only with notarial property site clauses and, to a much lesser extent, subscription clauses.

In crafting site clauses the notaries of late medieval Marseille had at their disposal a diverse array of possible cartographic templates: streets, districts of various types, islands, vicinities, and landmarks. These were the templates used by their clients. We must assume that the clients of the notaries, when first asked by the notary to give property sites for the purposes of a property conveyance, chose to define these sites according to their own languages of space. The diversity typical of vernacular linguistic cartography, at times, managed to push through whatever standard form the notaries might have been developing, because the set of extant notarial site clauses includes every possible template and all manner of toponymic styles. Even in the mid-sixteenth century, all possible templates are represented in notarial site clauses. This long-standing diversity is itself evidence for the absence of any formal conventions that might have governed the cartography of property sites.

All the same, fourteenth-century notaries had a clear preference for streets, and often translated the cartographic terminology of their clients into the language of streets. In so doing, they were building a map based primarily on a vision of the city as a tracery of streets, a map that tended to exclude vicinities, informal districts, and landmarks as possible cartographic templates. Moreover, comparison across the fourteenth, fifteenth, and sixteenth centuries shows that notaries were ever more inclined to build site clauses out of streets, alleys, and plazas. A wholly consistent street-based map, it is true, took several centuries more to emerge in Marseille. Nonetheless, the standardizing process was already underway in late medieval notarial practice.

If, as I have postulated, cartographic science in late medieval Marseille was in no way created or shaped by any centralizing forces or guiding legal, intellectual, or scientific schemes, how do we explain this slow transformation in notarial linguistic cartography? As I will argue below, the linguistic techniques developed to identify the spatial coordinates of property sites emerged naturally out of the ever-increasing number of cartographic conversations that took place between notaries and clients whenever property changed hands. These cartographic conversations, which resulted in coordinates that were set down on paper, inadvertently exposed the diversity of mental maps. Because the notaries, by the nature of their profession, participated in far more conversations than did others, the conversational space of cartography was, to a large extent, a notarial space; hence, any cartography preferred by notaries would be more likely to emerge as the basic linguistic convention. Medieval Marseille's notaries were especially conscious of streets, perhaps owing to the itinerant nature of their business. Thus, streets grew to be ever more common in notarial cartography, gradually crowding out the other possible cartographic templates. When a universal language of cartographic discourse finally emerged, by the seventeenth or eighteenth centuries, it became immediately available for other purposes. Among other things, it could be used by the record-keeping bureaucracies being developed by European states as a way of locating people in space. The address became part and parcel of the ideal world of the early modern state in which citizens are attached to addresses and do not move around inconveniently.

NOTARIAL CARTOGRAPHY

In locating urban buildings and other sites, the notaries of medieval Marseille dressed up, in Roman legal form, a technique for locating property characteristic of legal practice throughout much of the middle ages in many regions of Europe. This was simply to name the general location of the property and then to list whatever landmarks touched the building on its sides, usually four in number, but sometimes more or less depending on the shape of the building and the nature of the site and the adjoining properties.[6] The description of the property's location was embedded in the body of a notarial act in the "property site clause," derived from the word *sita* or *situm* ("situated"), the word that normally introduces this clause in Marseille. The property site clause in a contract was always followed by an abutment clause that described adjoining landmarks, introduced, in Marseille, by the word *confrontatum* or its equivalents. Together, these site descriptions form what would be called a plat description in contemporary U.S. administrative language, in this case a plat description that identifies a property site linguistically rather than graphically. House sales are the most frequent. They account for about a third of all contracts involving a conveyance of urban property in Marseille. This is so because it was relatively common for a wide spectrum of people, rich and poor, to be proprietors. Other acts with site clauses include sales of other kinds of urban property (such as gardens, building lots, and workshops), short-term leases, emphyteutic conveyances,[7] donations, and a wide range of other acts by which rights in property were transferred. For convenience, I shall call these collectively "property conveyances," bearing in mind that in some of the acts, such as testaments and dotal acts, the conveyance of property itself is but a small part of the purpose of the act.

Below is the protocol of a house sale. The two personal identity clauses, the property site clause, the abutment clause, and the subscription clause are in italics:

---

[6] David Herlihy, *Medieval Households* (Cambridge, Mass., 1985); Derek Keene, *Survey of Medieval Winchester*, 2 vols. (Oxford, 1985), 1:37.

[7] These are contracts in which a lord conveys a property to a proprietor in exchange for a single payment (known in Marseille as the *accaptum*) and a yearly rent (*census*), the amount of which could vary according to the size of the *accaptum*.

In the year of our Lord 1353, the next to last day of the month of October, around midday, let it be known to all etc. that *lord Raymon Lhaugier, priest, citizen of the city of Marseille,* in good faith and without any trickery or fraud, for himself and his future heirs, purely, truly, perfectly, and irrevocably transferred by title of sale and conveyed purely, freely, and absolutely without any expressed reservations to *Guilhem Tollaysi and Johaneta, a married couple, citizens of Marseille,* present, who for themselves and their heirs are purchasing, agreeing, and receiving, *a certain house situated in the city of Marseille in the street of St. Martin,* from the floor to the roof, with all rights and obligations. *It abuts on one side the house of Raymon Borier, and on another side the house of the heirs of the gentleman Guis and the house of Symon de Sancto Marcello, and in front the public street and behind another public street.* It is subject to the direct lordship of the monastery of the Blessed Mary of Ybelina of Porta Gallica of Marseille for a rent of 40 royal shillings payable each year on the feast of St. Thomas the Apostle. The price is 110 royal pounds, including the entry fee. The seller acknowledges having and receiving this price. Renouncing etc. If indeed it is worth more etc. Giving, ceding, and mandating etc. Constituting etc. Constituting etc. Remitting etc. and not transferring etc. Promising about eviction of tenants. Remitting etc. Permitting etc. Obliging etc. Renouncing etc. Swearing etc. Of these things the buyers requested that a public instrument be made. *Transacted in Marseille in a certain home of the seller near the Jewish fountain.* Witnesses Bernart de Casabona, Olivier Bonpar, Jacme Salvester. An instrument was made.[8]

Other property site clauses illustrating different topographical nomenclatures follow:

Example 1: *quadam domum sitam in Malocohinac del Temple* (a certain house located in the [vicinity of] Malcohinat of the Temple).

Example 2: *ante valvas inferioras ecclesie beate Marie de Acuis* (before the lower doors of the church of Notre Dame des Accoules).

Example 3: *ad cantonum Triparie* (at the corner of the Tripery).

Example 4: *ante Fontem Judaycum* (before the Jewish fountain).

Example 5: *retro domum Bondavini* (behind the house of Bondavin).

Example 6: *in carterio Sancti Johannis in carreria Figueria* (in the quarter of St. Jean in the street of the Figgery).

Example 7: *in carreriam Bernardi Gasqui ad partem bodii carreriam Botoneriorum* (in the street of Bernat Gasc; to the rear, the street of the Buttoners).

[8] ADBR 381E 79, fols. 100r–v, 30 Oct. 1353.

Example 8: *in insula Bertrandi Montanee* (in the island of Bertran Montanhe).

Example 9: *in burgum de Syon* (in the burg of Syon).

Between 1337 and 1362, sixteen active public notaries in Marseille produced seventy-two casebooks and registers of extensos, and the thousands of acts found in these registers contain 932 distinct property conveyances with legible site clauses.[9] One hundred and twelve clauses were composed of two or three terms (see examples 6 and 7 above) but for the sake of simplicity I will focus on the words or phrases that make up the first term or are in the first position in the site clause (Table 2.1).[10] For ease of analysis I have excluded expressions referring to the city of Marseille, to the three lesser cities, and to the "suburbs" (*surburbiis*), a generic term used in notarial records to refer to the ill-defined political jurisdiction that lay outside the walls.

Some of the terms in these site clauses consist of a type of space—street, island, or burg—and a toponym, as in examples 7, 8, and 9. Other terms are toponyms standing alone, as in example 1, in which the type of space, in this case a vicinity, must be inferred. Last, some terms consist of a relational preposition other than "in" or "at" coupled with a toponym, as in examples 4 and 5; this is the pattern typical of landmarks. The number of possible terms is not infinite, although the diversity is nonetheless impressive, especially in light of the much more uniform addressing template typical of modern bureaucratic practice and popular usage. The terms tend to fall into easily recognizable types, sometimes because the type of space is actually mentioned in the term and sometimes because the toponym is embedded within a distinctive grammatical structure.[11]

---

[9] Many property conveyances are often followed immediately by acts related to the conveyance that also contain site clauses. These include acts whereby the property lord approved the conveyance (*laudationes*) and acts whereby the conveyance was ratified by a third party with an interest in the property in question (*ratificationes*); this person was typically a woman who had dotal rights in the property in question. When these acts occur in series, the site clauses are repeated. I have counted these series as constituting a single site clause.

[10] Terms in second or third positions include fewer streets (44 percent), slightly fewer vicinities (12 percent), the same number of districts (17 percent), and more landmarks (27 percent). Typically, these landmarks were performing the role of an abutment, delimiting the specific location of the property on a given street, as in the example of a clause identifying a property site in "the street of the Workshops of the Tanners, behind the church of St. Louis"; see ADBR 355E 292, fols. 11r–12v, 24 May 1356.

[11] Naturally, there are a certain number of problematic terms—a place like the Jewish fountain, for example, looks like a landmark (a fountain), but functions grammatically more like a vicinity.

Table 2.1. Cartographic terms and categories used in notarial site clauses, 1337–62

	Number of times used[a]	%
Streets		
Street (*carreria*)	521	
Alley (*transversia*)	21	
Plaza (*platea*)	1	
Total streets	543	58.3
Districts		
Burg (*suburbium, burgum*)	72	
Unofficial district	51	
Jewry (*Jusataria*)	13	
Island (*insula*)	8	
Lesser city alone	7	
Administrative quarter (*sixain, carterius*)	4	
Parish (*parochium*)	0	
Total districts	155	16.6
Vicinities		
Artisanal or retail vicinity	86	
Landmark vicinity	67	
Total vicinities	153	16.4
Landmarks		
Church or hospital (*ecclesia, hospital*)	43	
Other	15	
Gate (*portale*)	10	
Notable man or woman	6	
Quay (*rippa portus*)	4	
Oven (*furnum*)	3	
Total landmarks	81	8.7
Total	932	100

Sources: ADBR 300E 6; 351E 2–5, 24, 641–45, 647; 355E 1–12, 34–36, 285, 290–93; 381E 38–44, 59–61, 64bis, 72–87, 393–94; 391E 11–18; AM 1 II 42, 44, 57–61.
[a] This column identifies only those terms used in the first position in notarial site clauses.

In grouping these terms into four larger templates or categories—streets, all types of districts (including islands), vicinities, and landmarks—I have imposed a classifying scheme that would not have been recognized by contemporaries. These categories, however, are useful for purposes of analysis, for although streets, say, are different from alleys, and plazas more different still, nonetheless they are all open spaces in which someone could walk, and they do not have boundaries. I have labeled the entire category "streets" simply because streets are by far the most important type of space in the category. Districts are larger than open spaces; they have boundaries, they include buildings, and they typically touch and incorporate several streets. I have included several different templates in this category, including administrative quarters, burgs, and islands, so as not to multiply the number of categories. It is important to bear in mind the distinctions between the elements of the category. Vicinities differ from streets because they incorporate a cluster of houses and can encompass more than one segment of a street; unlike districts, they do not have boundaries. The category of landmark is perhaps the most difficult to justify, because in many respects landmarks are used in contemporary records in just the same way that vicinities are used. Nonetheless, it is useful to distinguish them, if for no other reason than that the grammatical structures according to which they are referred are often distinct from those typical of vicinities.

Each of these types of spaces and categories deserves careful examination; each offers important lessons about the ways in which notaries, as cartographers, created and used the verbal map of the city. But first we need to ask the following question: Was this really a *notarial* cartography? Is there any reason to think it differed in significant ways from vernacular linguistic cartography? After all, notaries worked for clients who brought them sets of facts and asked them to arrange those facts in a legally binding form. Perhaps notarial cartography is a simple reflection of common usage.

By definition, we cannot hope to reconstruct the vernacular linguistic cartography of late medieval Marseille from notarial sources, and since the majority of extant sources were written by notaries, it is difficult to know where to turn. As it happens, however, among the few documents from medieval Marseille not written by a notary there is one that reveals how members of the non-noble Christian population—artisans, merchants, retailers, service trades, professionals, laborers, and fishermen— were inclined to map the city. This is a register of accounts kept, in Provençal, by one of Marseille's confraternities, the confraternity of St. Jacques de Gallicia, between the years 1349 and 1353. It includes long lists identifying the current members of the confraternity, and these personal identity clauses often include the members' places of domicile. This document is a subject in its own right, one I will take up in a later chapter.[12] For the moment, it is important only to note what the register

---

[12] See chapter 4 below.

reveals: members of the confraternity collectively referred to streets and other open spaces in only 13.3 percent of their addresses (Table 4.2), compared to the 58.3 percent we find in the case of notarized site clauses. They were correspondingly far more inclined to speak of artisanal, retail, or landmark vicinities, in 54.3 percent of their addresses compared to the 16.4 percent typical of notarial site clauses. In terms of patterns of usage, the Latin notarial cartography of site clauses neatly reversed the priorities of the vernacular linguistic cartography commonly used as addresses by the men and women of this confraternity. The comparison shows that the cartography of notarial site clauses was a distinctly notarial cartography. In creating site clauses, notaries were, to a great extent, translating the vernacular cartography given to them by their clients into a cartography that suited the norms of Marseille's notarial community. This was a cartography based on streets.

Streets

Use of streets was by no means the norm in site clauses and addresses across fourteenth-century Europe. As mentioned earlier, although streets were common enough in English, French, and Florentine records, other cartographic units, notably parishes and administrative districts, show up as the dominant template in a variety of other locales, especially on the Continent. When the notaries of fourteenth-century Marseille contemplated the map of the city, however, they normally saw streets. Streets alone account for over half of all terms used and are found six times more often than the next most common terms, burgs and artisanal or retail vicinities.

For all that, the map made by the notaries was not consistently a street-based map. Notaries themselves could alternate between templates. For example, the notary Peire Giraut used "the street of the Spur" in three site clauses but referred to "the Spur," using the vicinal template, in a fourth.[13] Moreover, notaries had not developed a standard nomenclature on the question of suburban usage, and used landmarks chosen on an ad hoc basis with some frequency. Most important, there were variations in style among Marseille's public notaries, some being more inclined than others to use vicinities, some more inclined to use landmarks, as a comparison of the four best-represented notaries from the mid-fourteenth century reveals (Table 2.2). Among other things, given notar-

---

[13] ADBR 381E 77, fols. 39v–41v, 11 May 1348 (this one refers to *Speronum*); 381E 78, fols. 86v–87r, 25 May 1350; 381E 81, fols. 131v–132r, 5 Mar. 1359; 381E 83, fol. 15r, 22 Apr. 1361.

Table 2.2. Comparison of cartographic categories used by members of two notarial lineages, 1337–62

	N	Streets (%)	Districts (%)	Vicinities (%)	Landmarks (%)
Giraut lineage					
Paul Giraut, 1337–46	69	42.0	18.8	27.5	11.6
Peire Giraut, 1343–62	237	59.1	18.1	16.9	5.9
Total	306	55.2	18.3	19.3	7.2
Aycart lineage					
Jacme Aycart, 1348–62	263	73.4	7.6	10.6	8.4
Peire Aycart, 1349–62	95	72.6	7.4	10.5	9.5
Total	358	73.2	7.5	10.6	8.7
Total	664	64.9	12.5	14.6	8.0

Sources: ADBR 391E 11–18; AM 1 II 59–60, ADBR 381E 76–83; ADBR 355E 1–12; ADBR 355E 34–36, 209–93.

ial lineages may have had distinct cartographic preferences. Peire Giraut, for example, was the son of Paul Giraut, and both were more inclined to use districts and vicinities and correspondingly less likely to use streets than either Jacme or Peire Aycart. Although Jacme and Peire may not have been closely related, both evince almost identical patterns of usage, even though Jacme lived and worked within the walls of the city, whereas Peire, for much of his career, lived and worked in the suburbs, an area whose map was quite different from the map typical of the intramural city.

The best method for predicting the cartographic usage of a given notary is age. Paul, who was especially inclined to forego streets in favor of districts (chiefly burgs) or vicinities (such as the Jewish fountain and the Pilings), was active as a public notary as early as 1318, whereas the other three notaries were first licensed in the 1340s.[14] This pattern also holds in the case of the father-son tandem of Bertomieu de Salinis and Johan de Salinis. Bertomieu, the elder, used vicinities in 41 percent of his forty-six site clauses; streets show up in only 22 percent. Johan, his son, used vicinities more commonly than other notaries—they show up

[14] Peire Giraut was licensed in 1342, Jacme Aycart in 1344, and Peire Aycart in 1348; see AM FF 166, fols. 6v–11v for a list of licensed notaries drawn up in the year 1351. Paul Giraut was already dead by this point, so we do not know when he first began practicing. His first extant casebook, however, is from the year 1318; see ADBR 391E 1.

in 34 percent of his twenty-nine site clauses. Yet streets become much more significant in Johan's casebooks, since they are used in 45 percent of his site clauses.[15] This generational change typifies the historical drift toward the nearly universal street-based urban map of early modern Marseille. The Black Death may have played a role in this change, too, although exactly what role is hard to tell. Most of the older notaries— Paul Giraut, Bertomieu de Salinis, and Bernat Blancart—died in the Black Death or shortly before. Augier Aycart, who was licensed on 4 May 1304, and was perhaps the oldest of them all, lived until 1352. The notaries who came into their own after the Black Death—Peire Giraut, Jacme Aycart, Peire Aycart—were increasingly inclined to use streets in their site clauses.

All elements of the city's landscape were capable of being transformed grammatically into streets, simply by writing the toponym in the genitive and attaching the word *carreria* or its equivalents. As we have already seen, the Spur could appear as a vicinity (*ad Speronum*), but could be changed with relative ease into the street of the Spur (*in carreria Speroni*). As a result, we cannot tell *a priori* how many streets mentioned in notarial documents were widely accepted components of the vernacular linguistic cartography. In particular, we do not know whether other people called them "streets," and we cannot be sure that the toponym used by a notary was a widely accepted toponym.

A total of 179 different streets, alleys, and plazas show up in notarial property conveyances between 1337 and 1362. Like streets in medieval cities in the kingdom of France, they took their names from a wide variety of toponyms.[16] Some toponyms, about 26 percent, were based on the city's physical landmarks, such as churches, monuments, gates, ovens, fountains, and baths. Others, 19 percent of the total, took their names from occupational groups, markets, and other centers of retailing. Another 19 percent bore names of uncertain provenance whose etymologies are not obvious, names that were not tied in any clear way to the city's physical, social, artisanal, or commercial structures. Examples include the New street, perhaps the product of some earlier architectural re-engineering; the street of the Strangers (*carreria Foresteriorum*), possibly named because it was located near an area where inns were grouped; and the street of Cina, which bore a name of uncertain origin.

The most frequent toponyms were those derived from notable men or women or families; these account for about 36 percent of the total. Some

---

[15] See ADBR 381E 38–44; AM 1 II 57–58, ADBR 381E 72–75.

[16] See the categories of toponyms described in Jean-Pierre Leguay, *La rue au moyen âge* (Rennes, 1984), 94–98.

toponyms were conventional; that is to say, they typically took a family surname alone, and were not necessarily tied to a living individual or family. The prosopographical index tells us, for example, that no one bearing the surname Johan lived on the street of the Johan between 1337 and 1362. Other streets took the forename and surname of a specific individual, such as the street of Guilhem Folco, named after a prominent merchant of the mid-fourteenth century; typically, such toponyms changed with the passing of the person in question, although some could become conventional and endure for a few more generations.[17] In some cases, we can witness the process whereby a family name became a conventional feature of the city's linguistic cartography. Such is the case with the prominent noble family of the Grifen, many of whose members lived on a street in the *sixain* of St. Martin, running parallel to the street of St. Martin. In various notarial site clauses from the mid-fourteenth century, the street was once called "the street of Guilhem Grifen"; once it was named after Guilhem's brother (or son) Nicolau Grifen; but on four occasions it was reduced to the more convenient "the street of the Grifen."[18]

These streets did not all have the same status. Some were conventional elements of notarial toponymy, like Negrel street, the New street, the street of the Tannery, the street of the Jerusalem, the street of Raymon Rascas, the street of St. Martin, the street of the Smiths, and many others. They show up in a number of different site clauses in the casebooks of different notaries and are usually spelled the same way. Negrel street was one exception; it sometimes appears as the *carreria Negrelli* and sometimes as the *carreria de Negrel* or *de Negrello*. Nonetheless, the existence of orthographic norms is an important clue; it tells us that these streets and many others like them were widely recognized as relatively stable elements of notarial linguistic cartography. These names were occasionally written in Provençal. A street named after Madam Capona was usually called, in Latin, the street of Capona (*carreria Capone*), but sometimes appears as the *carreria de Na Capona* or the *carreria dan Capona*. A long, well-known street called, in Latin, the *carreria Francigena*, once shows up in Provençal as the *carreria Franceza*.[19] Some of these streets evidently figured in vernacular linguistic cartography as well.

---

[17] Colin Platt discusses interesting examples of changing toponyms that reflect the changing prominence of local families in *Medieval Southampton: The Port and Trading Community, A.D. 1000–1600* (London, 1973), 47–48.

[18] See ADBR 381E 80, fols. 13r–v, 31 Dec. 1354 (*carreria Guillelmi Griffedi*); ADBR 381E 83, fols. 39r–41r, 23 May 1361 (*carreria Nicolai Griffedi*). One example of the usage of *carreria Griffenorum* is ADBR 381E 82, fols. 40v–41r, 6 May 1359.

[19] ADBR 355E 6, fols. 44r–46r, 22 July 1353. This is in a site clause in the testament of a woman named Alazays Arnosa.

Elsewhere, the street map was a little more uncertain. In a dotal act from 1358, the notary Peire Giraut recorded a house site as "the street of Guilhem Folco or the Spur," reflecting either his own uncertainty or that of his clients about the name of the street in question.[20] Sometimes this uncertainty is reflected in expressions that included the words "called" (*vocata*) or "named" (*dicta*) so-and-so, as in the case where the notary Jacme Aycart identified a house located in "a street that has been called *dan Dieulosal* for a long time" (*carreria dicta ab antiquo dan Dieulosal*).[21] Despite this assurance of antiquity and hence conventionality, no other notary ever used that toponym in mid-fourteenth-century Marseille. Notaries clearly made up some streets as they went along, for example by transforming landmarks into streets, as in the case of "the street behind St. Marthe above the corner" or "the street of the Quay of St. Jean," both of which were used only once.[22] Other unique names, like "the street of Marita Vivauda," "the street of Peire Milhayrole," or "the street of Bernat Girone," seem to have been created in an ad hoc fashion from the name of a notable woman or man.[23] It is easy to imagine the situation: the client, asked to give the property site, says "it's across the street from Marita Vivauda" or "it's near Peire Milhayrole's house," and the notary, not familiar with the site in question, turns the personal name into a street. A total of thirty-eight streets were used only once in notarial site clauses between 1337 and 1362; twenty-eight of these were named after people. Many of these unique toponyms may have been ad hoc inventions.

Not having access to the conversations between client and notary that lay behind the making of these site clauses, we have no way of measuring the degree of conventionality of a given toponym. The street of Uguo Pedagier appeared only once in site clauses between 1337 and 1362. This may reflect the lack of conventionality of the name, but may also be due to the stability of the proprietors who lived on the street. It is even possible that a certain notary, whose casebooks are now lost, monopolized business on the street. The fact that most of these streets do not show up in any other documents from the period does tend to suggest the ephemeral nature of the names they bore. And in point of fact it is easy to appreciate why some unconventional names would show up in site clauses. Notaries, after all, did not have an official cadaster to which they could refer, nor is it likely that every notary kept the map of the entire city in

---

[20] ADBR 381E 81, fol. 33v, 8 July 1358.
[21] ADBR 355E 12, fols. 38v–41r, 8 June 1362.
[22] ADBR 351E 647, fol. 156r, 21 Sept. 1357; ADBR 381E 77, fols. 139r–v, 15 Oct. 1348.
[23] ADBR 381E 77, fol. 133v, 30 Feb. 1349; ADBR 381E 80, fol. 29v, 26 Jun. 1354; ADBR 381E 76, fols. 71v–72r, 1 Feb. 1348.

his head. Most property conveyances were not written *in situ*, and this distancing in the act of conveyancing added to the definitional problems.[24] In constructing site clauses, therefore, notaries had to rely to a certain extent on information brought to them by their clients. Unique street names, in many cases, may simply reflect the particular cartographic lexicons of local communities, lexicons not shared by other people or, for the most part, by the city's public notariate. If this is true, then it is arguable that notaries, or at least the property conveyances they supervised, played an important role in mediating between the cartographic lexicons and styles used by different communities of knowledge in the city, thus helping to create a universal lexicon.

City and Lesser Cities

In mid-fourteenth-century Marseille, public notaries were by no means translating all property sites brought to them by their clients into the language of streets. There is another way of looking at the relative dominance of streets in fourteenth-century notarial cartography. After all, 38 percent of notarial site clauses never made mention of streets in either first or second positions, and it is worth looking at the other possibilities. Starting at the highest possible level, many but no means all site clauses make reference to the fact that the house in question was located in the city of Marseille, *in ville Massilie.* In a few site clauses, seven in all, one of the lesser cities was the only term in the site clause. In 1343, for example, the notary Augier Aycart supervised the sale of four building lots (*casalia*) located simply "in the upper city," abutting on the back side a house belonging to the squire Isnart Beroart, a house belonging to the almoner of St. Victor, a garden belonging to Raymon Julian, and an unnamed public street.[25] Such cases reveal that site clauses did not necessarily have to be precise; it was the abutments that did the work of identifying the precise location.

In most clauses that mentioned one of the lesser cities, the primary site would nest within the city; an example might read "located in the lower city of Marseille in the New street." The notaries of Marseille were licensed to transact business outside the confines of the city and its territory and

---

[24] In the subscription clauses in 230 house sales between 1337 and 1362, only five (2 percent) took place at the house in question. Other locations were as follows: lord's house, 67; house of seller or seller's husband, 52; house of unrelated person or other neutral space, 41; notary's house or shop, 35; court, 17; buyer's house, 13. In a few cases it is possible that the seller's house or buyer's house was located near the house being sold.

[25] ABBR 381E 393, fol. 116r, 23 Jan. 1343.

on rare occasions did indeed supervise property conveyances in nearby towns and villages, but so exceptional were these cases that many notaries took the larger city for granted. There was a marked disinterest in mentioning a lesser city, namely, the upper or lower cities or the Prévôté, even before the formal union of the cities late in January of 1348. The upper and lower cities were wholly distinct legal entities, with distinct jurisdictions and independent councils. Relations between the cities were not always cordial. Even citizenship, before 1348, was tied to one or the other of the two lesser cities.[26] Yet people circulated freely through both cities, and notaries practiced indiscriminately in both. Possibly as a result of this freedom of practice, notarial cartography does not reflect legal boundaries: in thirty-two sales of intramural houses from 1337 to February of 1348, we find only four mentions of the lower city and two of the upper city, or in 19 percent of the sales, and the site clauses in other contracts reveal a similar percentage. Lesser cities, moreover, were used most often by older notaries like Augier Aycart, Paul Giraut, and especially Bernat Blancart.[27] Their fitful use of the lesser cities reflects a fading memory of an age in the thirteenth century when the two cities had more distinct political and cultural identities. All references to lesser cities in contracts of sale disappear after February of 1348, and the last reference I have found to one of the lesser cities in notarial sources, in the emphyteutic conveyance of a house, is dated 16 November 1349.[28] The new legal status of the city surely played some role in this change, but the death of older notaries—the plague played a large role in this—also served to silence the cartography of the older generation.

Burgs

Outside the walls of Marseille, extending in an arc from the northern apex of the city to the southeastern corner, were the suburbs or burgs. An active market in suburban property produced a total of 200 notarial site clauses involving suburban property between 1337 and 1362. Sometimes the suburbs were called just that, "the suburbs," suggesting that they could be imagined as a single political entity on a par with the three intramural cities; a similar expression was "outside the walls" (*extra muros*).

[26] Two acts of citizenship, both pertaining to the upper city, have survived in the notarial records extant between 1337 and 1362; see AM 1 II 60, fol. 14r, 4 Dec. 1343, and idem, fols. 6r–v, 19 Mar. 1344.
[27] A fourth notary who used the sub-cities on occasion was Johan de Salinis, a notary of considerable standing and long experience in the profession.
[28] ADBR 355E 2, fols. 122r–v, 16 Nov. 1349.

Table 2.3. Cartographic categories used in 200 notarial site clauses concerning suburban property, 1337–62

	Number of times used	%
Streets	109	54.5
Districts	72	36.0
Vicinities	5	2.5
Landmarks	14	7.0
Total	200	100

Sources: ADBR 300E 6; 351E 2–5, 24, 641–45, 647; 335E 1–12, 34–36, 285, 290–93; 381E 38–44, 59–61, 64bis, 72–87, 393–94; 391E 11–18; AM 1 II 42, 44, 57–61.

Most site clauses, accordingly, were based on streets nested within this generic entity. The notary Paul Giraut, for example, gave one property site as "in the suburbs of Marseille in the street of the Oven of Peire Austria" (*in suburbiis Massilie in carreria furni Petri Austrie*); his son Peire Giraut identified another as "the street of the Tannery in the suburbs" (*carreria Blancarie in suburbiis*).[29] As in the case of the lesser cities, notarial site clauses did not always mention the larger entity, sometimes taking it for granted. In analyzing suburban site clauses, I have not included references to this generic entity (Table 2.3).

A total of twenty-eight different streets and two named alleys show up in notarial site clauses involving suburban property from 1337 to 1362, sometimes alone or embedded in "the suburbs," sometimes in combination with landmarks or specifically named burgs. Ten streets took their names from one of the churches or mendicant establishments that had been founded in the suburbs in the thirteenth century, and another street was named after a landmark, an oven. Eleven were based on names whose origins are not evident. Four streets were named after notable individuals, three of whom were women: Jauma Bellauda, Madam Capona, and Madam Auriola. The tendency to mention women in suburban toponymy is interesting; by way of comparison, the intramural city, with six times as many named streets and alleys, produced only four streets named after women: Adalays de Monacabus, Madam Pescayressa, Madam Gazanha, and Marita Vivauda. Four streets in the suburbs were named after artisanal groups or artisanal toponymy: tanners, corders, the Locksmithery, and the Tilery.

The most distinctive feature of suburban linguistic cartography was the custom of calling some areas "burgs." Notarial site clauses between 1337

[29] ADBR 391E 15, fols. 39r–v, 5 July 1341; ADBR 381E 78, fols. 96v–97r, 7 June 1350.

and 1362 commonly made reference to nine; clockwise from the north, these were the burgs of Aygadiers, Oliers, Moriers, St. Clare, Prat d'Auquier, Syon, St. Augustine, Dominicans, and St. Catherine, situated to the south and west of the city. Two others show up more rarely: Robaud (one site clause) and Malamortis (two site clauses). Prat d'Auquier also shows up in five site clauses without the word "burg"; here, as elsewhere, it takes the grammatical form typical of a vicinity.

Despite the frequency of usage, the cartography of these burgs is not obvious. Among other things, their political, administrative, and fiscal status was not comparable to that of the *sixains,* and perhaps for this reason they did not have borders—at least, borders defined in an administrative sense. To judge at least by mid-fourteenth-century notarial site clauses, some of the burgs, namely those of Aygadiers, Moriers, St. Clare, Prat d'Auquier, and St. Catherine, contained no more than a single street that bore the same name. In these cases, using the name of the burg in the site clause was essentially the same thing as using the name of the street. In contrast to this, the burg of the Dominicans, the burg of the Augustinians, and the burg of Syon included at least three streets and alleys within their borders, and the burg of Oliers was composed of at least two streets, the street of Oliers and the street of Jauma Bellauda. We know this because site clauses referring to houses in these burgs mentioned both burg and street with some frequency, as in a clause in a donation drawn up by Johan Silvester in 1351 identifying a house located "in the street of Bocharts in the burg of St. Augustine."[30] Other contemporary documents allow us to identify even more streets considered to be within the borders of the three largest burgs, Syon, St. Augustine, and Dominicans.

One result of this is that site clauses concerning these burgs often mention not only the name of the burg but also (and often exclusively) the name of a street within the burg. The burg of the Dominicans is especially interesting in this regard. Among the fifty-three notarial site clauses identifying property located within the burg of the Dominicans, we find forty-three sites identified by street or alley alone (e.g. "in the street of the Pilas"), five identified by both the burg and the street ("in the street of the Pilas in the burg of the Dominicans"), three landmarks, and only two that used the burg alone ("in the burg of the Dominicans"). In the case of the other burgs, the practice of avoiding the use of burg alone was not so marked. The "burg of St. Augustine" appears alone in five of the thirty-one site clauses concerning property located in the burg; in the

---

[30] ADBR 358E 84, fols. 63v–64r, 29 July 1351.

case of the burg of Syon, the ratio is six in eighteen. Notaries tended to dissect burgs into streets when possible or necessary, either by using a street-based cartography or by coupling streets with burgs.

Yet if all these areas were mappable by means of streets, why did notaries bother to mention the burgs at all? They did so because their clients did so, and vernacular cartography, here as elsewhere, pushed through the notarial preference for streets. Burgs—or *borcs*, as they were sometimes known in Provençal—show up frequently in the addresses found in seigneurial, judicial, and even notarial records from the period. For example, in the 376 addresses found in the register of the confraternity of St. Jacques de Gallicia from 1349 to 1353, six burgs—Oliers, Moriers, Syon, St. Augustine, Dominicans, and Jarret—were used a total of twenty-seven times. When residents of the suburbs did not use the language of burgs, moreover, they preferred to use landmarks and not streets.[31] In a record of episcopal rents from 1343 to 1347, the 324 addresses include twenty-two references to eight burgs and only five suburban streets.[32] To judge by these examples, vernacular linguistic cartography typically mapped the suburbs by means of the language of burgs. Vernacular cartography used all the names of burgs found in notarial site clauses and, on a few occasions, seven additional ones, the burgs of Bocharts, Jarret, Franciscans, Madam Auriola, Madam Capona, Pilas, and Raymon Rascas. In the context of cartographic imagination, for Provençal speakers these burgs were probably very similar to the vicinities typically found within the walls of the city, that is to say, knots of sociability with ill-defined borders that stretched over one or more streets and included the houses. This sociability is attested by the fact that four burgs—Syon, Dominicans, St. Augustine, and Moriers—were associated with confraternities or "lumenaries" that bore the name of the burg; it is not surprising that these four burgs regularly show up in addresses.[33]

One of the most interesting features of notarial suburban cartography is that a given notary was more likely to use the vernacular language of burgs in his site clauses, and more likely to using nesting site clauses, if he himself was generally unfamiliar with suburban topography. We cannot

---

[31] See Appendix 1.

[32] ADBR 5G 114.

[33] For the "lumenary of the burgs of St. Augustine and the Dominicans," see ADBR 381E 83, fols. 35r–37r, 21 May 1361; for the "lumenary of the burg of the Dominicans" (probably the same as above), see ADBR 355E 293, fols. 53r–56v, 24 March 1363; for the "lumenary of the burg of Syon," see ADBR 355E 35, fols. 1r–v, 26 March 1357; this lumenary was also called "the lumenary of St. Bernard of the burg of Syon"; see ADBR 358E 84, fols. 34r–36v, 25 May 1351; for the "confraternity of the lumenary of the burg of Moriers," see ADBR 381E 80, 73v–76r, 27 Sept. 1354. These are all notarized testaments.

measure the degree of notarial familiarity with suburban cartography, of
course, but we do know that comparatively few notaries lived in the
suburbs. Of a total of seventy-nine notaries between 1337 and 1362 whose
likely place of domicile is known, only seven lived in the suburbs (Table
2.6); of those whose casebooks have survived, only one was a suburban
resident. For most notaries, their suburban practice was rather limited.
This is not because they didn't do business with a suburban clientele; they
did. But their suburban clients, who were typically poorer, usually came
to them, not the other way around. The notary Peire Giraut, for example,
lived on the street of St. Martin close by one of the gates that led out into
the burgs of Syon and St. Augustine. Peire conducted about 75 percent
of his business outside his home. Yet in a sample of fifty-five acts that
involved clients who lived in the suburbs, we find that he traveled the few
hundred yards to the suburbs in eighteen instances, only 33 percent of
the total. In forty-eight sales of suburban houses in all notarial casebooks
between 1337 and 1362 in which the site of the transaction can be
identified, only four took place in the suburbs, primarily because sellers
and lords, those who most commonly determined the transaction site,
were typically wealthy citizens and nobles who lived within the city walls.
Distance was not an obstacle. Virtually the entire urban area of medieval
Marseille, suburbs included, was located within a thousand meters of the
plaza of Accoules at the heart of the city.

   One result of these preferences is that notaries were less familiar with
the suburban map than they were with the intramural map. This, at any
rate, is what we can deduce from the fact that the one notary represented
by extant casebooks who *did* live in the suburbs, Peire Aycart, was far more
inclined than his colleagues to use the language of streets alone, and far
less inclined to refer to burgs alone. Only four of his thirty suburban site
clauses were based on burg alone; one mentioned "Prat d'Auquier" as a
vicinity; and the remaining twenty-five were all based on streets. His colleagues were also more prone to nest streets within burgs, in a sense to
combine vernacular and notarial cartographies. This is not to say that
other notaries did not use streets elsewhere in Marseille. It simply shows
that the more familiar the notary was with a given site, the more likely he
was to impose his cartographic preference for streets on the site clauses
provided to him by his clients.

Administrative Quarters and Other Districts

   As noted earlier, the lower city was divided into six administrative districts called *sixains* (St. Jean, Accoules, Draparia, St. Jacques, Callada, and

St. Martin) and the upper city into four quarters (Rocabarbola, St. Cannat, Annonaria, and St. Jacques of the Swords), and the language of *sixain* and quarter was the primary cartographic language of the city councils of both cities. One of the most curious features of the linguistic cartographies of medieval Marseille is that these entities almost never occur in property site clauses, identity clauses, or subscription clauses. Like parishes, they simply did not figure in either notarial or vernacular cartography. The only major exception is the sixain of St. Jean, which crops up ten times in acts drawn up by three different notaries, on four occasions as the first term in the clause and on six occasions as the second term. Seven terms make explicit reference to the "quarter of St. Jean" (*carterio Sancti Johannis*); in the other three cases, the usage suggests that the notary had the *sixain* in mind, and not the church of St. Jean.[34] Six cases involved houses located on the street of the Figgery, and this helps explain in part the recourse to the *sixain*: a street with the identical name was located in the suburbs, and another could be found in the upper city. Given the existence of three Figgeries, the identification of the *sixain* helped avoid confusion. The confusion was real: on 27 November 1349, the notary Jacme Aycart, in writing out an act of sale, described the house as situated "in the street of the Figgery in the burg [elided] the quarter of Sancti Johannis" (*carreria de la Figayressa in burgo* [elided] *carterio Sancti Johannis*).[35]

Two other usages of "the quarter of St. Jean" seem gratuitous; one embraced a wholly unambiguous street, the street of the Toesco, inside the quarter of St. Jean, and another site clause read "in the quarter of St. Jean, near the quay."[36] Here, we cannot explain the use of the *sixain* as a function of the need for cartographic precision. In point of fact, fishermen and mariners did commonly speak of the quarter of St. Jean; they used it, for example, in giving their addresses, as in the case where the fisherman Guilhem Martin gave his address as "the quarter of St. Jean" to the notary Jacme Aycart in an act in 1359.[37] In a register of episcopal rents from 1365, moreover, this same Guilhem Martin shows up again, and again he used the quarter of St. Jean as his address. It was evidently something on which he insisted.[38] What this suggests is that

---

[34] The use of *carterio* or quarter instead of *seyzeno* or *sixain* is surprising: the *sixains* of the lower city were never called "quarters" in other records I have examined.

[35] ADBR 355E 10, fols. 83v–85r, 27 Nov. 1359.

[36] ADBR 355E 8, fols. 59r–61r, 8 Jan. 1356; ADBR 355E 12, fols. 156r–157r, 8 Oct. 1362.

[37] ADBR 355E 10, fol. 36v, 21 July 1359. See also 355E 9, fols. 75r–v, 18 Sept. 1358; 355E 11, fol. 8v, 17 Jan. 1361; and idem, fol. 23v, 11 Mar. 1361, for other examples.

[38] ADBR 5G 116, fol. 3r.

the quarter of St. Jean, unlike the other *sixains*, had become a conventional residential space for members of seafaring trades, something
with which they identified. It was an identification not so much with
the concept of the *sixain* as with a shared space or habitat. This may
explain why they always used the word "quarter" (*quarterius, carterius*) to
refer to the district and never the administrative language of *sixain*
(*seyzenum*).

As the example of St. Jean and the burgs suggests, vernacular linguistic cartography spontaneously invented large districts that bore little or
no relationship to administrative districts. The most spectacular example
is the unofficial district of the upper city known as Cavalhon, used alone
in fifteen notarial site clauses and in combination with islands, streets,
or landmarks in a further nine. As was the case with the burgs, the geographic limits of this district are difficult to pin down. Cavalhon seems to
have encompassed every street and island west of the street of the Upper
Grain Market, north of the lower city, and east of the Prévôté. It also gave
its name to a street, which shows up seven times in notarial site clauses.
Another large district is that of the Upper Grain Market, which appears
ten times in site clauses unaccompanied by the words "street" or "quarter."
The pattern of usage and the consistency in orthography very strongly
suggest that when a notary or his clients used the expression, they did not
mean the grain market itself (*annonaria*), nor the street (*carreria Annonarie
Superioris*), nor the quarter. Nineteen site clauses refer to Rocabarbola,
the name of an administrative quarter of the upper city. As in the case of
the Upper Grain Market, I think it is unlikely that notaries or their clients
meant the administrative quarter when they used the name. Among other
things, neither notaries nor their clients ever used the names of the two
remaining quarters of the upper city, St. Cannat and St. Jacques of the
Swords. Complementing Cavalhon, Rocabarbola seems to have been used
commonly to refer to a triangle that had the street of the Upper Grain
Market and the lower city as its western and southern borders and the
city ramparts to the north and east.

Whatever notaries and clients had in mind when speaking of these districts, the cartography of the upper city was unique in its frequent use of
these three large-scale districts. This unique cartography reflects, at least
indirectly, the cartographic hegemony of the episcopal curia, the former
banal lord and still major direct lord of the upper city. Episcopal insis-
tence on using islands as the basic cartographic term in rent registers
seems to have hindered the linguistic development of named streets, and,
uninterested in the cartography of islands, ordinary speakers of the ver-

nacular turned instead to these districts.[39] The districts were the upper-city complement to the geographically smaller vicinities that are so marked a feature of the vernacular cartography of the lower city.

Jewry

For medieval notaries, as well as modern historians, perhaps the most mysterious space of all was the Jewish quarter, the Jewry. In the map accompanying Octave Teissier's *Marseille au moyen âge*, the Jewry is presented as a large space criss-crossed by several unnamed streets which were clearly hopeful attempts on the part of Teissier to draw what he assumed must have been there.[40] Bruno Roberty's map is equally revealing. A more careful scholar, he left the Jewry largely blank and only drew in a few well-documented streets in the northeastern corner as well as two intersections, one coming off the New street to the south and the other cutting in on the western edge (Figure 3).[41] The enormous cartographic uncertainty reflected in both these maps is a fair reflection of what the archives have left us in terms of a good understanding of the northeastern corner of the Jewry, though little else.

The reasons for this are interesting, and involve more than an aesthetics of disappearance.[42] Houses located within the Jewry were identified in fifteen notarial acts; the Jewry was specifically mentioned in thirteen. In seven of these, the district constituted the entire site description (*domum sitam in Jusataria*), and in another six the site—the well-documented northeastern corner—was further delimited by reference to streets or alleys, the nesting pattern occasionally used in suburban site clauses. One of the remaining two conveyances of property in the northeastern corner of the Jewry named a street, and the other a notable person.

From these acts we learn that the northeastern corner, located near the church of St. Martin in the general neighborhood of Marseille's cosmopolitan Jewish property lord and moneylender, Bondavin, was interpenetrated by a large Christian population.[43] This Christian presence, and the acts that result, made the space knowable by Christian notaries.

[39] See chapter 3 below.

[40] Octave Teissier, *Marseille au moyen âge. Institutions municipales, topographies, plan de restitution de la ville* (Marseille, 1891).

[41] ADBR 22F 140.

[42] On this see Kathleen Biddick, "Paper Jews: Inscription/Ethnicity/ Ethnography," *Art Bulletin* 78 (1996): 594–99.

[43] Bondavin, one of Marseille's most prominent citizens, figures in Joseph R. Shatzmiller's *Shylock Reconsidered: Jews, Moneylending, and Medieval Society* (Berkeley, 1990).

Significantly, only one of the fifteen site clauses comes from an act involving a transfer between two Jewish parties. In this case two women, Pebra
Mossa and her sister, Belleta Bondia, were selling a house to the Jewish
physician, Salves de Cortezono, a syndic of Marseille's Jewish community.
It is unlikely, to say the least, that this single act represents all the property conveyances between Jews between 1337 and 1362, and hence we
are forced to conclude that many property conveyances between Jews
were not handled by Christian notaries, and that notaries were only called
in to deal with conveyances in the Jewry involving Christian clients.[44] With
the Christian property-owning population limited to the northeastern
corner, there are simply no extant acts from which we might reconstruct
the cartographic lexicon of the remainder of the Jewry. The seven acts
referring loosely to the Jewry may well have involved properties located
in the mysterious inner part of the Jewry, but if so, this was not an area
whose streets were known to notaries.

Islands

With the "island"—in Latin the *insula*, in Provençal the *illa* or *isla*—we
return to the upper city, and this time at a lower geographic level than
before. This is not because there were no entities called islands elsewhere
in the city—there were, and lots of them—but instead because the upper
city was the only political jurisdiction in which a notary made reference
to them.

An island is a city block, that is to say an irregular parcel of land
bounded on three or more sides by streets or, in some instances, by a city
wall. They are common in registers of rental obligations kept by officials
of both the episcopal curia and the Angevin curia in the mid-fourteenth
century; these records were almost invariably organized according to
island. The Angevin crown possessed rents (*censuales*) on several hundred

---

[44] In twelfth-century Marseille, Jews had their own legal devices for the conveyancing of
property; see the formularies of Isaac Ben Abba Maris of Marseille written between the years
1172 and 1189, published in Hans-Georg von Mutius, ed., *Jüdische Urkundenformulare aus
Marseille in babylonisch-aramäischer Sprache*, vol. 50 of *Judentum und Umwelt* (Frankfurt, 1994).
These include formularies for the sale, purchase, and lease of real property (see 32–39).
For a discussion of Jewish scribes, see Shlomo H. Pick, "The Jewish Communities of Provence
before the Expulsion in 1306" (Ph.D. diss., Bar-Ilan University, 1996), 182–84. I am grateful to Dr. Pick for bringing this matter to my attention. It is exceedingly unlikely that
fourteenth-century property conveyances among Jews might have been recorded in Christian notarial casebooks that did not survive. The notary Peire Giraut, for example, lived very
near the Jusataria and wrote hundreds of contracts for his Jewish clients. If Jews had their
intra-community conveyances notarized, there is every reason to think Peire would have
been one of many notaries favored for this business.

houses situated in forty or fifty islands in the lower and upper cities. The record-keeping practices of the royal curia, however, almost never spilled over into the practice of the public notariate: notaries simply did not use royal islands in their site clauses. The cartographic style of the bishop's curia, in contrast, did have a life outside episcopal registers. Islands sometimes appear alone in the site clauses (eight cases; Table 2.1) and sometimes follow either streets (three cases) or the district of Cavalhon (six cases). These instances of insular usage come primarily from the casebooks of a single notary, Esteve Venaissin.

Why did Esteve's fellow notaries almost never use islands in their site clauses? Islands are no less accurate a means for identifying property sites. Consider the house purchased by Raymon Blanc from Berengiera Assauda in February of 1354.[45] Esteve, the notary who drew up the act of sale, noted that the house was located "in the island of den Feroge, abutting a house belonging to Johan Tomas, the orchard of Johan de Quinsac, and the garden of Guilhem Porcel." As this example shows, site clauses within the island template used here by Esteve included abutments that carefully identified the location of the property on the given block with no possibility of ambiguity. This being so, the answer cannot lie in the superior ability of the notarial street-based method to identify houses. This is especially true when we consider that notaries were not at all reluctant to list a large and ungainly district like Cavalhon and Rocabarbola or even a lesser city itself as the only term in a site clause.

Nor can we explain the notarial tendency to avoid islands as a product of notarial ignorance of the insular template. For one thing, the people responsible for keeping both episcopal and royal registers were, invariably, notaries. Many of the notaries who wrote acts acknowledging episcopal lordship that appear in a register from 1343 to 1347, including Guilhem Tornator and, toward the end of the register, Rostahn Columbier, Johan Bermon, Guilhem de Belavila, Antoni Pascal, Raymon Vidaut, Johan de Thama, and Esteve Venaissin, were also active as public notaries. The latter two are represented by surviving casebooks.[46] The topography in this register was shaped by the insular template; hence, these notaries had to have been aware of the template. Furthermore, the activities of the episcopal and royal curias would have ensured that knowledge of the template circulated outside the narrow circles of these two major direct lords. Among other things, the acts drawn up on their behalf invariably required witnesses, often members of the general public who

---

[45] ADBR 351E 3, fols. 21r–v, 25 Feb. 1354.
[46] ADBR 5G 112.

would have had occasion to hear a verbal description of the location of the property. Thus, it is very likely that the insular template was widely known in Marseille.

To explain the general notarial reluctance to use the cartography of islands, it might be useful to rephrase the question and pose it this way: Why did Esteve Venaissin depart so markedly from the notarial norm? The answer is that, by and large, he did not.

Esteve worked frequently for the episcopal curia in the mid-fourteenth century, and therefore knew about and used the insular template in his work. He also kept a small practice as a public notary on the side, and like any notary he drew up numerous property conveyances.[47] These involved houses and gardens located not just in the upper city but also scattered throughout the lower city and the suburbs; in his acts we find houses located in the street of the Jewish fountain, in Gache street, in the Fishmongery, in the Furriery, before the house of Bernat de Casabona, in the Spur, in the street of the baths of Tholoneum, and so on. Only when the property site was located in the parts of the upper city domi-nated by the lordship of the bishop did Esteve use the language of islands, as in clauses like "the island of Bertran Montanee," "the island of Guilhem Andree," "the island of Bertran Lombart," and such nesting site clauses as "in Cavalhon, in the island of Johan Lavandier." What we seem to have here, then, is a case where a notary, familiar with episcopal cartography and accustomed to thinking of a certain part of the city in terms of islands, naturally used islands in his relevant site clauses.

But the situation is more interesting than this, for Esteve switched from one template to another even within the upper city itself. We find clauses referring to the district of Rocabarbola, Francigena street, the Upper Grain Market, and "near St. Marthe." Even in areas of the upper city that fell under episcopal lordship, Esteve used non-insular templates from time to time, such as the street of St. Jacques of the Swords, the street of Pons Broquier, the Gentleman's Oven, "behind St. Antoine," and the large district of Cavalhon. All these sites could be and were mapped as islands in records from the episcopal curia. Yet Esteve, or his clients, chose to avoid episcopal cartography on certain occasions.

Remarkably, his manner of use was to a large extent a function of where the conveyancing act was written. Between 1351 and 1362, Esteve wrote twenty-six conveyances identifying thirty-two properties in the upper city. Twenty-one of these houses, identified in eighteen acts, fell under the lordship of the bishop. From the subscription clauses to these acts, we can

---

[47] For his casebooks, see ADBR 351E 3–5. I also used two extensos: ADBR 351E 642 and 351E 645.

tell that all but three took place within the confines of the episcopal compound (*actum in domo episcopalis*).[48] This was unusual, for most direct lords took less interest in the physical space of the conveyancing act itself and were more likely to control the location of the acknowledgment of lordship made by the new owner that followed. For reasons that are not evident, the episcopal curia had an interest in supervising conveyances of property under episcopal lordship.

This supervision had an interesting effect on the way in which Esteve drew up his site clauses. When Esteve wrote a contract in the episcopal compound, he generally used the insular template favored by the bishop; we find thirteen insular usages as the first or second term in a total of eighteen site clauses drawn up in the compound. However, when Esteve wrote a contract in a place removed from direct episcopal supervision, on three of four occasions he described the site simply as Cavalhon. The evidence is small but the difference suggestive: outside the confines of the episcopal compound, Esteve was inclined to revert to a more normal notarial or vernacular cartography. He *could* have chosen to use an insular template in these three cases, because he surely knew the names of the islands on which the properties were located. Instead, he preferred to use the template offered him by his clients. Cavalhon, however large and ungainly a district, was a cartographic term favored both by the residents of the upper city and by public notaries.

Esteve was not alone in feeling the influence of the episcopal curia. The only other notary to refer to islands in house sales from the mid-fourteenth century was Peire Giraut. Peire, who lived in the lower city, had far less contact with the episcopal curia than did Esteve; his "extracurricular" work was typically consecrated to service in one of the courts of law in the lower city. Yet in one house sale, from June of 1350, the property site clause he drew up identified a house located "in the street of Jacme Ancel, in the island called Paul Naulon's" (*in carreria Jacobi Ancelli, in insula dicta Pauli Nauloni*).[49] The lord of the property was the bishop; the transaction took place in the episcopal compound. A few weeks later, Peire wrote up two more house sales in the episcopal compound, involving a house located "in the street of Pons Broquier, in the island den Lavandier" and another "in the street of Pons Broquier, in the island of Guilhem de Tolosa."[50] These are the only three property conveyances written by Peire involving property under episcopal lordship. All three were written in the episcopal compound. He used islands in all three. No other property conveyances drawn up by Peire in all his extant casebooks

---

[48] See, for example, ADBR 351E 3, fols. 21r–v, 25 Feb. 1354.
[49] ADBR 381E 78, fols. 108r–v, 30 June 1350.
[50] Ibid., fols. 119r–v, 18 July 1350; ibid., fols. 120r–v, 19 July 1350.

mentioned islands. This is a clear indication of the influence exerted by the episcopal curia on notarial cartography. Rather than ask why notaries avoided islands, then, it would be better to ask how and why the episcopal court, deliberately or accidentally, promoted the insular template, and in so doing managed to impose its wishes on public notaries.[51]

Vicinities

Vicinities were areas that could incorporate not only a street segment but also an intersection of streets and the alleys running off the streets as well. Between the usages of street and vicinity there is a very slight but telling grammatical distinction. In notarial casebooks, streets, alleys, and plazas almost always appear in the genitive in Latin or, in the rare instances when the notaries slipped into Provençal, with *de* or *del*. They are prefaced with the Latin words *in carreria*, *in transversia*, or *in platea*. Vicinities can be distinguished from streets, at least grammatically, because they appear in the accusative in Latin without any other words like *carreria*. In Provençal, they appear after such prepositional expressions as *a la*, *en*, or *en la*. Thus, in September of 1357 the carpenter Peire Bonfilh sold to the laborer Peire Garriga a house and garden located in a vicinity called "Malcohinat dal Temple" (near the old church of the Templars in the southeastern corner of the city).[52] "Malcohinat," from time to time, was also Latinized as a street (*carreria Malcohinati*). Other examples comparing Provençal and Latin grammatical structures follow:

Provençal vicinal format	Latin vicinal format	Latin street-based format
en la Pelisaria Estrecha	in Pellipariam Strictam	in carreria Pelliparie Stricte
en la Frucharia	in Fruchariam	in carreria Frucharie
en la Qurataria	in Curatariam	in carreria Curatarie
a la Pescaria	in Piscariam	in carreria Piscarie
al Forn Danprodome	ad Furnum Danprodome	in carreria Furni Danprodome

Although the street-based usage was most common, notaries did slip into a Latin vicinal format in 16.4 percent of the 932 property site descrip-

[51] The theme I take up in chapter 3.
[52] ADBR 355E 35, fols. 79r–v, 6 Sept. 1357.

tions from the mid-fourteenth century (Table 2.1). Since all vicinities could be and often were Latinized as streets, it is important to note that quarters with relatively homogenous populations of artisans or retailers, such as the Carpentery, the Goldsmithery, and the Cobblery, were somewhat more likely to appear grammatically as vicinities. By the same token in notarial site clauses, areas that had lost a craft identity were being transformed into streets. A case in point is the Smithery, where few smiths resided in the mid-fourteenth century. Notarial site clauses refer to the area on fourteen occasions. In every single clause, the area was called "the street of the Smiths" (*carreria Fabrorum*) and never "the Smithery" (*Fabraria*).

Interestingly, notaries were twice as likely to use the language of craft vicinities in site clauses in short-term property leases (*locationes*) of workshops or houses located in artisanal vicinities. In these cases the buildings being leased were places of work, and the lessees were members of the trade that dominated the vicinity. Thus, notaries were more willing to use what appears to have been the preferred nomenclature of their artisanal clients in cases where the craft-based interests of the clients so powerfully shaped the meaning of the act.

Other vicinities were not attached by their names to specific artisanal or retail centers, although neighborhood identities seem to have been just as powerful. Instead, they typically bore a name that was, in origin, a landmark, transformed by usage into a vicinity. In these cases it is not clear why the notaries chose to transliterate some as streets and leave others in the accusative, other than the simple fact that they were conforming to common usage and that the long-term shift from vicinal to street nomenclatures did not proceed evenly throughout the city. Of the dozens of such vicinities only a few, namely Panataria, Stone Image, Lancery, Malcohinat, Arches, Corregaria, Pilings, the Enclosure, Malausena, and the Butte of Rocabarbola, appear in the accusative in property site clauses; these also appear as streets from time to time. Others, such as Pretty Table, the Hill, Emmendats, the Auction, the Spur, Roiling Stone, and Quay appear only as streets in site clauses or else do not appear at all, either because no properties in these areas were transferred between 1337 and 1362 or, in some cases, because notarial cartography used an entirely different lexical term for the area.

Landmarks

The landmark was yet another common template used to provide a basic cartographic grammar for property sites. Like the vicinity, the landmark was used much more frequently by members of the city's artisanal, retail, professional, and laboring population in vernacular linguistic cartography. Within this template, sites were identified simply in relation to some nearby monument. Churches or hospitals, for example, are referred to on forty-three occasions in property conveyances. A notary might say that the house was located "above St. Esprit," "before Our Lady of Acuis," "behind St. Martin," or "near the cathedral church of Sedis in the street which leads up to the said church." In the case of the spaces surrounding churches, the notaries were often dealing with entities rather like plazas that had buildings in the middle. In other cases, gates, fountains, and ovens were used to define open areas that were too awkwardly shaped to be called either streets or plazas, so one finds descriptions like "in front of the pastry oven of the Fishmongery" and "at the gate of the Frache." Over time, these spaces acquired generic terms in both common and notarial usage. Landmarks were still in common usage as late as the mid-sixteenth century, although the types of imaginable landmarks were somewhat less diverse, limited largely to elements of the city's physical make-up, like churches, gates, towers, and the quay. With a few exceptions, ovens, markets, and notable men and women had been entirely eliminated from the toponymic repertoire of sixteenth-century notaries.

Churches, hospitals, and other civic buildings were favored landmarks even in the fourteenth century because they were large and enduring edifices, easy to find and hence ideal for navigation. Interestingly, churches located on street frontages rather than at the center of open spaces figured less often in the fourteenth-century landmark template. The church of St. Martin, for example, occupied what was more or less a discrete block (a cemetery was attached at the back); in a sense, it sat in the middle of a large square. The houses that faced the church could not easily be mapped by means of streets. As a result, it is common to find houses situated "near St. Martin," and the street of St. Martin refers not to the square in which the church was located, but instead to the major road that led to the square. The church of St. Marthe in the upper city, in contrast, was located on a street frontage with houses on both sides. Three properties located on the street in front of the church of St. Marthe in the upper city were identified not as being located "near St. Marthe" (*prope Sanctam Martham*); instead, all three were located in the "street of St. Marthe" (*in carreria Sancte Marthe*). Because the church of St. Marthe was visibly located on a street, its name had simply become associated with that street. Similarly, a house on the street in front of the seat of the crown in Marseille, the Palace, was located on the "street of the Palace" rather than "near the Palace" (*prope Palacium*); like St. Marthe, the Palace was located on a street frontage. As these examples suggest, notaries preferred street-based usage where possible, and only had recourse to the landmark template when dealing with open spaces that were hard to imagine as straight streets.

Subscription Clauses

Leaving behind property sites, we turn briefly to subscription clauses. These are intrinsically less useful than property site clauses because most transactions took place in one of a limited set of possible locations, such as the notary's home or workshop, the houses of one of the parties involved, one of Marseille's courts, or some other neutral site, such as a church. The clause often states "transacted in the house of so-and-so" (*actum in domo habitationis*) and little else. In 1139 loans and debts enacted between the years 1337 and 1362, something more than *actum in domo* was given in only thirty-two acts. The notaries invariably used *burgum* in preference to *carreria* when speaking of suburban transaction sites in seven cases, and used adjoining landmarks in two instances (*prope portus* and *iuxta vallata Sancti Spiritus*). In eleven cases where the notaries had a choice between using a vicinal form or a street form, the notaries chose the street in nine instances (e.g., *carreria Petrem Ymaginis, carreria Curatarie*) and the Provençal norm in only two (*ad Collam* and *Sabataria Scarie*). The remaining twelve cases were all streets, and thus streets were used in twenty-one out of thirty-two occasions or 66 percent of the time, comparable to notarial usage in site clauses. I have found no references to islands, quarters, parishes, or any large districts (except for suburbs) in any of the subscription clauses in all of the notarial acts I have studied from the period 1337 to 1362.

Making the Map

Over time, the street template became more and more normative in notarial practice. A fairly rapid and unscientific sampling of 269 site clauses drawn from the casebooks of twenty different notaries active between 1445 and 1455 reveals that usage of streets and similar open spaces had increased by some fifteen percentage points (Table 2.4). Usage of adjoining landmarks rose slightly, whereas usage of districts declined dramatically, partly because the depopulation of the city caused by the plague and by the great sack of Marseille in 1423 meant that the

Table 2.4. Comparison of cartographic categories used in notarial site clauses, 1337–1555

	N	Streets (%)	Districts (%)	Vicinities (%)	Landmarks (%)
1337–62	932	58.3	16.6	16.4	8.7
1445–55	269	72.9	4.5	10.8	11.9
1545–55	262	74.8	9.2	6.5	9.5

Sources: *Line 1* (1337–62): ADBR 300E 6; 351E 2–5, 24, 641–45, 647; 355E 1–12, 34–36, 285, 290–93; 381E 38–44, 59–61, 64bis, 72–87, 393–94; 391E 11–18; AM 1 II 42, 44, 57–61. *Line 2* (1445–55): ADBR 351E 285, 330, 333, 344, 367, 378, 379, 402, 408, 409, 433; 352E 7; 355E 108, 133, 142; 358E 7, 283; 373E 6, 15, 37; 381E 90, 107. *Line 3* (1545–55): ADBR 351E 876; 352E 133, 144, 148; 354E 36; 356E 16; 357E 25; 358E 18; 359E 55; 362E 3, 12; 363E 10, 12, 42; 364E 1; 366E 83; 373E 194.

suburbs had shrunk in size. But demographic changes alone cannot explain the most interesting observation that emerges from the comparison between fourteenth and fifteenth-century usage—usage of vicinities in the first position in notarial site clauses dropped by a third. This trend continued into the sixteenth century, for a similar survey shows that royal notaries from mid-sixteenth-century Marseille, seventy years after Provence had been incorporated into the kingdom of France, had nearly halved their frequency of usage of vicinities.

Streets did not gain noticeably in the aggregate between the fifteenth and sixteenth centuries, although in point of fact a more careful study would show that usage of streets was continuing to increase. Among other things, if we look at *all* parts of notarial site clauses, not just the first position, we find that reference to streets rises consistently across the centuries (Table 2.5). The apparent discrepancy is easily explained, given that in the sixteenth century, a number of notaries writing site clauses had fallen into the habit of writing more nesting clauses. In particular, they more commonly built clauses in which streets nested within quarters, with the quarter appearing as the first term. A second point to bear in mind is that variations in practice among notaries appear to have been even greater in the sixteenth century than in the fourteenth century. Some sixteenth-century notaries in Marseille, such as Bertrand Rebier, Jean Boutaric, Jean de Scalis, and Jean de Olliolis, collectively used streets in 93 percent of their site clauses; others, notably Henri Sicolle and Jean Gandelbert, used streets in 57 percent.[53] This suggests that the community of notaries in

[53] The figures are: Bertrand Rebier, 9 in 9; Jean Boutaric, 6 in 7; Jean de Scalis, 6 in 6; Jean de Olliolis, 30 in 33. These contrast sharply with Henri Sicolle, 15 in 24; Jean Gandelbert, 6 in 13. See, respectively, ADBR 363E 12, 362E 10, 366E 83, 373E 194, 359E 55, and 363E 42.

Table 2.5. Streets in notarial site clauses, 1337–1555

	N	One or more streets[a] (%)	No streets (%)
1337–62	932	62.0	38.0
1445–55	269	75.5	24.5
1545–55	262	82.8	17.2

Sources: Line 1 (1337–62): ADBR 300E 6; 351E 2–5, 24, 641–645, 647; 355E 1–12, 34–36, 285, 290–293; 381E 38–44, 59–61, 64bis, 72–87, 393–394; 391E 11–18; AM 1 II 42, 44, 57–61. *Line 2* (1445–55): ADBR 351E 285, 330, 333, 344, 367, 378, 379, 402, 408, 409, 433; 352E 7; 355E 108, 133, 142; 358E 7, 283; 373E 6, 15, 37; 381E 90, 107. *Line 3* (1545–55): ADBR 351E 876; 352E 133, 144, 148; 354E 36; 356E 16; 357E 25; 358E 18; 359E 55; 362E 3, 12; 363E 10, 12, 42; 364E 1; 366E 83; 373E 194.
[a] Column 3 gives the percentage of clauses that mention streets in the first or in subsequent positions.

mid-sixteenth-century Marseille was, for whatever reason, less cohesive. The marked gap in the cartographic styles of notaries might have had something to do with the abandonment of Latin and the taking up of French as the language of record. Sixteenth-century notaries, all of whom wrote in French, may have been trained elsewhere in France and may have been importing a distinctively French linguistic cartography from these regions into Marseille. In this light it is interesting to note that Henri Sicolle changed habits later in his career, using ten streets in fourteen site clauses found in a register of extensos between 1570 and 1594.[54] Perhaps it took him that long to absorb the cartographic style of Marseille's notariate.

As the comparison shows, the act of making the map was not only a matter of translating vicinities and landmarks into streets. It was also a historical process that resulted in an ever more consistent and standardized map of the city, the street-based map that, soon after, began appearing in graphic representations of the city. Here, then, we arrive at the fundamental question of why notarial site clauses favored streets as the basic unit of cartographic awareness. One possible exogenous agent of change was the guiding influence of the *ius commune*. One could argue that Marseille's notaries were simply conforming to some legal norm of cartographic precision imposed by the *ius commune* and that the transformation

[54] ADBR 359E 57.

from the common cartographic usage of Provençal speakers to Latin notarial cartography was the result of the penetration of Roman law into everyday practice. Yet the answer is not easy as this.

First, the legal formulas of the notaries demanded a property site clause, but these formulas—to judge, at least, by the very diversity of possible plat descriptions in the mid-fourteenth century—did not specify the cartographic template used in the clause. In theory, the phrase "located in Marseille" could have served the legal formula quite well, and indeed site clauses were not infrequently as vague as this. The geographical designation that figures in notarial formularies from the first half of the thirteenth century, in fact, is the parish (*parrochia*) and not the street; hence, Marseille's notariate in the fourteenth century was departing from Italian notarial custom.[55]

Second, site clauses rarely mattered in courts of law. In the several thousand court cases I have read from the fourteenth century, I have seen a site clause referred to only on one occasion.[56] They were never used to establish title in a house. Courts of law in fourteenth-century Marseille used written documents primarily for the purpose of verifying credits and debts. When faced with a dispute over ownership of a house or other piece of property, they looked for evidence from neighbors that ownership had been publicly established by means of entering and exiting the house; the ownership and use of keys; and payments of both ground rents and taxes. Contracts of sale for houses were used primarily to establish the existence of a debt and to guard against fraudulent sales by men or women on the brink of insolvency.[57] They were not deeds. In a curious way, then, the actual house and its location were peripheral to these more significant debt-related concerns. It did not matter precisely how a site was identified so long as the form of the law was followed. This being so, notarial cartography did not need to convey some legal definition of a plat, and instead reflected the way in which notaries or their clients typically thought about space and location.

Last is a simple but absolutely fundamental point: as argued above, the vicinal and insular templates were every bit as precise, in their legal description of a property site, as the street template. When appear-

---

[55] For formularies using parish (*parrochia*) see Salatiele, *Ars notarie*, ed. Gianfranco Orlandelli, 2 vols. (Milan, 1961), 2:229, and Bencivenne, *Ars notarie*, ed. Giovanni Bronzino (Bologna, 1965), 38.

[56] See ADBR 3B 862, fols. 161r–243r, case opened 19 Jan. 1409. The case concerns a direct lord trying to evict a proprietor from his house for failure to acknowledge the lord's rights.

[57] Daniel Lord Smail, "Notaries, Courts, and the Legal Culture of Late Medieval Marseille," in *Urban and Rural Communities*, ed. Reyerson and Drendel, 48–49.

ing in property conveyances or rent registers, all three templates listed a site—street, island, or vicinity by implication—and all used abutment clauses to delimit the precise location of the property, the abutment clauses serving much the same function as street numbers do today. Streets were not the unique solution to some legal demand for precision in the verbal representation of space; the law alone cannot explain why notaries preferred a street template.

To explain notarial preference for streets, we need to look within notarial culture. Vernacular nomenclature could carry a social meaning, notably in the case of artisanal or retail vicinities where the name reflected, or hoped to reflect, a certain homogeneity of livelihood and a resulting vicinal solidarity. The implied sociability of vicinities and craft group quarters was centered on an open space and joined in face-to-face relations people related to one another by bonds of friendship, if not professional ties. It was a usage that carved up the city into introspective, non-communicating, corporate segments, even if in practice vicinities did not always achieve this goal.

Notaries did not live in a notarial vicinity. Although many lived in the *sixain* of Accoules, near the center of commercial and legal activity, their residences were distributed widely across Accoules itself, and others lived elsewhere in the city. Among those whose casebooks have survived, there is Augier Aycart, in Accoules, above St. Esprit; Bernat Blancart, in Accoules, in the Coopery; Raymon Audebert, in Accoules, near the Palace; Jacme Aycart, in the street of Perier, in the upper city just above the church of Notre Dame des Accoules; Peire Aycart in the burg of Syon until 1359, when he moved to the street of the Tanners in the *sixain* of Callada; Paul Giraut and his son Peire, in St. Martin, near the Jewish fountain; Guilhem Johan, in the upper city, in the street of the Well of Velans; Bertomieu de Salinis, in Accoules, in the street of the Servenha; Johan Silvester, in Accoules, in Negrel street; and Esteve Venaissin, in the upper city, near St. Marthe, on the street of the Upper Grain Market. The prosopographical index provides the likely place of residence of seventy-nine notaries.[58] The evenness of residential distribution is impressive (Table 2.6).

Although notaries were distributed broadly throughout the city, they nonetheless had a diverse clientele, and as peddlers of the law, they typi-

---

[58] In some cases the prosopographical index lists only house ownership. Although it is likely that the notary actually lived in the house in question, it is not certain. Similarly, notaries like Peire Aycart were moving into the lower city from suburbs and the upper city throughout the 1350s; in many cases, the index reflects this change. Table 2.4 lists the first known (or likely) place of residence.

Table 2.6. Social topography of Marseille's notaries, 1337–62

Sixain or District	
Accoules	18
Cavalhon	15
Draparia	8
St. Jacques	7
Rocabarbola	7
Suburbs	7
St. Jean	6
St. Martin	6
Callada	5
Total	79

Source: Prosopographical index.

cally brought the law to their clients, not the other way around. We can actually measure this mobility, since the notary wrote down in the subscription clause just where the act was made. To take one example, the notary Peire Giraut enacted more than three-quarters of his acts (1100 out of a total of 1450) away from his home office and street, either in the houses of his clients or in the law courts. In these perambulations, they had to walk down streets, and if we try to imagine city space as they saw it, we can easily imagine that their mobility made them more inclined to *see* streets. They did not see islands, because to fully apprehend an island one has to walk around it (as seigneurial officials apparently did from time to time). Nor were they inclined to see vicinities, because vicinities were places in which one lived and practiced a trade, not places one walked through.

Moreover, one can speculate that a vicinity with close, neighborly bonds and a strong identity would have had less need for notarial services. Notaries, after all, relied on income derived from transactions made between people who did not know each other very well, certainly not well enough to trust either in an oral guarantee, or in the suasion offered by a patron, or in the constraints imposed by the culture of honor and shame. The notaries, instead, provided people with more readily negotiable access to a very different form of power—the courts of law. It is exceedingly difficult to assess how often loans and other favors passed between friends, neighbors, and close kin; in other words, it is not easy to document just how Marseille's vicinities might have functioned as social units. What we do know is that many types of notarial contracts— loans between Christians and Jews are a case in point—define a contractual engagement in writing between two people belonging to different communities of knowledge. To borrow Armando Petrucci's felicitious term, a notary was "someone who knows."[59] Notaries mediated between communities of knowledge, competing with both patrons and artisanal groups to act as brokers, providing their clients with access to resources. According to this hypothesis, notaries would have a vested interest in dismantling any structures that would challenge their mediating role. Thus, it is in the very nature of their trade that one can begin to see why notaries may not have been sympathetic to the vicinity as a tool of carto-graphic description. This argument is necessarily speculative, and will be borne out or disproven only when we have enough comparative studies on which to base a larger study of notarial cartography in medieval Europe.

Was this a conscious antipathy? There is no reason to think so. Had the notaries been fully aware of the reasons for their street-based cartogra-phy, the transformation would have proceeded with greater rapidity. What seems more likely is that a slight notarial preference for streets, embed-ded in the ways in which notaries apprehended the world, transmuted over the centuries into a systematic norm of linguistic cartography. To explain this process we must return to the law. The practice of law brought about a change in cartographic understanding not through its formal rules but rather because the property site clauses demanded by legal acts of conveyancing forced both the notaries and their clients to create a lan-guage for something they simply may not have thought about a great deal before, namely the identification of space. It is a question, therefore, of conversation.

Imagine, if you will, the following scenario, based on a genuine nota-rized house sale from the year 1350 and several other documents relative to the case. Late in the spring of that year, a banker and nobleman named Raymon del Olm agrees to sell to Resens Bernarda, wife of the laborer Guilhem, part of a house located in the southeastern part of the *sixain* of Callada. Raymon is the direct lord of the portion of the house he is selling to Resens; the other part of the house falls under the direct lordship of the crown. On 2 May 1350, Resens and Raymon meet the notary, Peire Giraut, who is to draw up the act of sale in the house of the shoemaker, Bertran Cuende, located near the house being sold.[60] Raymon brings along a friend of his, the esquire Marques de Jerusalem, and there is

[59] Armando Petrucci, "Pouvoir de l'écriture, pouvoir sur l'écriture dans la renaissance ital-ienne," *Annales ESC* 43 (1988): 834.
[60] ADBR 381E 78, fol. 71v–72r, 2 May 1350.

another witness, Johan Jausaqui, to round out the little party of six who officially participate in the transaction.

Peire, the notary, sets to work. Arriving at the property site clause, he pauses to ask those present to define the location of the house. Perhaps Raymon del Olm, inclined to see the city in terms of noble families, explains to him that the house is right around the corner, just above the street of the Johan. Perhaps Resens tells him that it is located next to the gate of Lauret. Bertran Cuende is the only member of the party who actually lives in the area, and he may just say that the house is in "Malcohinat" (in fact, this is how he gave his address to the scribe of the confraternity of St. Jacques de Gallicia).[61] Peire, the notary, lives north of the area, in the *sixain* of St. Martin; perhaps he has always found this area a little confusing, toponymically speaking. It abuts an area called the Lower Tannery, although the tanners have long since abandoned the area. He knows that Malcohinat alone is ambiguous, because there is another area called Malcohinat in the *sixain* of St. Jean. He may also know that the crown, one of the direct lords to whom the property owed rent, customarily identifies the site as the island of Guilhem d'Acre, even though Guilhem has been dead for some time and royal officials have begun to call it the island of Gautelm Malet.[62] In any event, no royal officials are present at the conveyancing and are not constraining the notary in any way.

This reconstruction only approximates the actual conversation, for all we have access to are the results. In the contract of sale, the notary Peire Giraut wrote down that the house was located "in the street of Malcohinat Lauret, near the street of the Johan," abutting a house belonging to Peire Guilhem, the street of Malcohinat, and the street of the Johan. When Raymon del Olm got home and had time to record the transaction in his own cartulary, several weeks later, he wrote a slightly different site clause, one that made no mention of streets: "Malcohinat del Lauret, above the corner of the Johan."[63]

The buyer, Resens, and the shoemaker, Bertran Cuende, may have been experiencing this sort of conversation for the first or second time; over their lifetimes, they might participate as buyers, sellers, or witnesses at a dozen such conveyances. Raymon del Olm was a property lord, and therefore his experience with the cartography of site clauses would have been greater. But none of them could match the cartographic mastery of the

---

[61] This is how he describes his address in 2HD E7, p. 41.
[62] See ADBR B831, fol. 44r for a description of the island.
[63] See ADBR 1HD H3, fol. 35r.

notary, Peire Giraut. Like any notary of the fourteenth and fifteenth centuries, Peire talked about property locations several times a month and
had hundreds of cartographic conversations over the course of his career.
As a notary, he dealt not with the same houses or gardens year after year,
like a property lord, but with parcels of urban property scattered all over
the city. From his casebooks we learn that Peire supervised conveyances
concerning urban property in twelve of the burgs and in all districts of
the upper and lower cities. His site clauses identify eighty-six different
streets and alleys, close to half the total of named streets. These repeated
conversations made Peire an expert in urban cartography, and it is an
expertise he would have shared not only with his clerks but also with his
fellow notaries, as he witnessed their acts and they witnessed his, and as
they discussed their professional attitudes at guild meetings. Notaries such
as Peire became, in effect, the official cartographers of the city, and the
template they developed, by its growing clarity, gradually became the city-
wide template, replacing those used by ordinary citizens and by the great
lords. The law, then, was responsible for this development only insofar as
it promoted cartographic conversations. The results of those conversations were shaped by notaries who, for reasons based partly in the way
they imagined urban space, tended to prefer a cartography based on city
streets.

Cartographic conversations were not particular to Marseille; they were
taking place wherever scribes or notaries drew up contracts related to the
conveyance of rights in property, in the north as well as in the south of
Europe. They were becoming increasingly common across the fourteenth
and fifteenth centuries, as property holdings, especially after the Black
Death, were turning over more rapidly, spurred by vacancies and urban
immigration. Based on the example of Marseille, this template may have
become the bureaucratic norm in many places of western Europe by the
end of the fifteenth century and the first decades of the sixteenth. There
was surely a great deal of diversity in the way this process unfolded in different regions of Europe, and just as surely no straightforward trajectory.
Derek Keene notes that in medieval Winchester, properties were listed
under street headings in rentals and account rolls, and parishes came to
replace streets as the primary site description in the fifteenth century, a
tendency at odds with the trend in Marseille.[64] Elsewhere, notaries may
not have been the only agents of change, as in the case of Italian cities

---

[64] Keene, *Survey*, 1:127. Keene also notes that "In London, where individual streets were
less readily identifiable with particular neighbourhoods, parishes as well as streets were generally mentioned in deeds from the thirteenth century onwards." Ibid.

where a universalizing process in linguistic cartography may have been equally associated with communal interest.

A street-based cartography was not a unique solution to the problem of mapping urban space; in subsequent chapters we will look more closely at things only touched upon here, in particular, some of the templates used by other linguistic communities. The problem itself had only emerged in the centuries after 1100 or so, the period of rapid urban growth in Europe and property turnover, and it is therefore not surprising that, even in the mid-fourteenth century, numerous cartographic templates jostled against one another. Even though they preferred a street template to any other, the notaries of the mid-fourteenth century themselves were eminently pragmatic, and sought only to draw up reasonably accurate property site clauses. Clearly the clauses drawn up by notaries reflected common usage to some extent, especially in the chosen toponyms. Like their clients, notaries avoided administrative and insular templates, and they also used some of the categories preferred by their clients. But they frequently departed from common usage in their strong preference for streets, translating the oral descriptions of property sites provided by their clients into the street template. In this way they acted, however unwittingly, as cartographers, that is to say, professionals who not only absorbed a cartographic awareness, but also digested and subsequently regenerated it. In what they created, notaries were adhering to well-known scientific processes of making heterogeneous entities standard and interchangeable. As an institution they were, over the centuries, engaging in an act of classification. Indeed, they were doing so before large-scale state sponsorship of graphic cartography and even before there was state interest in any form of cartography whether verbal or graphic.

Did this process of classification have a subsequent life? We know that in the early modern era, graphic representations of urban space from an aerial perspective have become commonplace. Part of the novelty of these representations, in comparison to their medieval antecedents, was an awareness of streets. We have no fourteenth-century representations of Marseille. To judge by the nearly contemporary paintings of Giotto and Lorenzetti, artists viewed their cities from a ground-based perspective that precluded the visual reconstruction of streets; a similar horizontal perspective is typical of non-western cities as well. The act of representation focused on buildings in elevation, stacked on top of one another as if there were no streets in between, as well as on the activity of people on the level of the street. Paintings often focused on noble towers, identifying the city partially with its magnates. The city was not the tracery of its streets but was rather its citizens, monuments, and sacred sites.[65]

As streets grew straighter and more prominent in early modern maps and artistic renderings of cities, the buildings that dominated medieval urban representations shrank in proportion and became quite uniform before disappearing altogether. The sixteenth-century engraving "Marssilia" (Figure 4) is typical of the urban vistas produced in that century. The urban zone consists of clusters of stylized houses nestled within a highly schematic tracery of streets. Other prominent elements closely associated with the identity of the city also stand out, such as the chain across the port, the windmills, the churches of Notre Dame des Accoules and St. Antoine, and the walls. "Marssilia," of course, was produced at a time when notaries had long since achieved a fairly uniform street-based cartography. The medievalizing nature of the image underscores the fact that the standardization process in cartography was linguistic long before it was graphic. A century or more later, however, streets have penetrated graphical consciousness. "Marseille, ville considerable de Provence" (Figure 5), an image from 1702, has abandoned all houses and nearly all monuments. What leaps out of the image is an accurate tracery of streets, an effect heightened by the engraver's decision to shade in the city blocks. Taken to the extreme, this tight association between streets and urban identity produced, in the second half of the eighteenth century, an image of Marseille and its coastline in which the city itself is symbolized by the skeletal form of its major thoroughfares (Figure 6). The gradual elimination of houses and monuments, as represented by these images, was a dehumanizing process. Trends in graphic cartography conspired to eliminate all reference to people in the urban imagination.[66]

This is not to say that the late medieval notariate uniquely influenced these transformations in artistic and cadastral representations.[67] It is only to suggest that the late medieval notariate contributed to this development by providing a universal cartographic language. More studies of the notariates of other cities will be needed before we can generalize freely.

---

[65] Particularly helpful on these issues is Chiara Frugoni, *A Distant City: Images of Urban Experience in the Medieval World*, trans. William McCuaig (Princeton, 1991).

[66] On these issues see J. B. Harley, "Silences and Secrecy: The Hidden Agenda of Cartography in Early Modern Europe," *Imago Mundi* 40 (1988): 65–66.

[67] The use of linear perspective in art, for example, contributed to changes in the ways in which cities were represented in art; see Samuel Y. Edgerton, Jr., *The Renaissance Rediscovery of Linear Perspective* (New York, 1975).

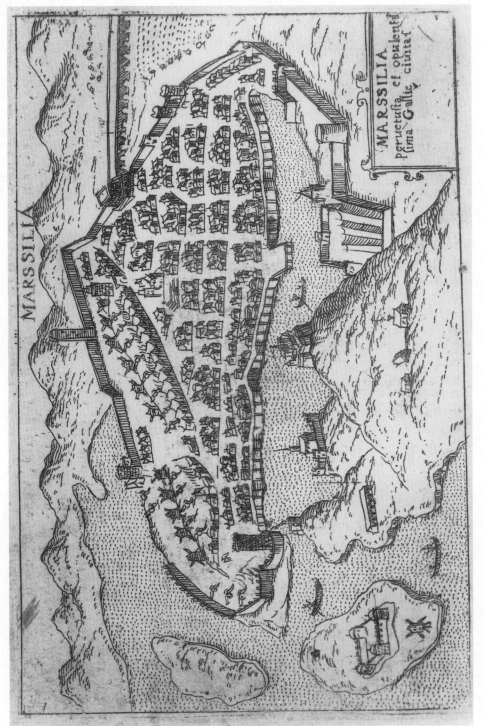

4. "Marssilia," late sixteenth century.

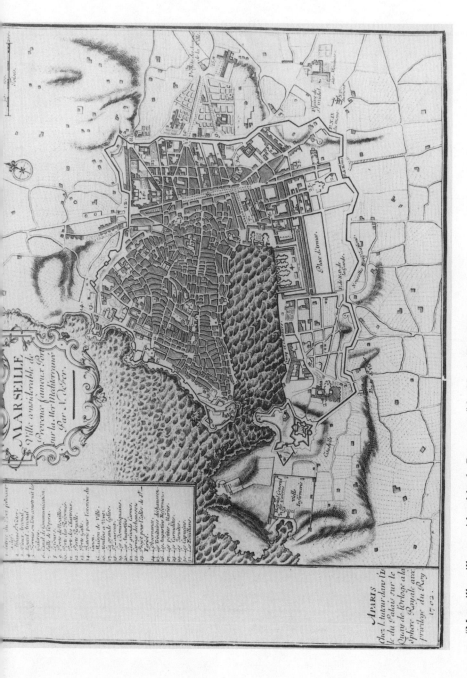

5. "Marseille, ville considerable de Provence," by N. de Fer, 1702.

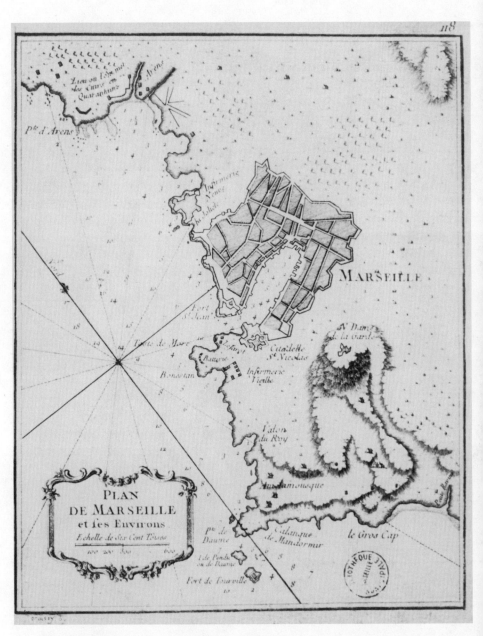

6. "Plan de Marseille et ses environs," attributed to Croisey, ca. 1750–1800.

The evidence from late medieval Marseille does suggest that the timing of the emergence of aerial viewpoints is not a coincidence.

To conclude, a street-based graphic model of the city was simply not conceivable until the cartographic debates of the later middle ages had been resolved. These debates or conversations took place on those occasions when notaries, clients, and lords got together to discuss a conveyance of rights in property and, in so doing, necessarily discussed the description of a property site. These conversations were only necessary after the emergence of the public notariate ensured that property conveyances would, in the future, be written down, complete with site clauses and other cartographic apparatus. In other words, the growing popularity of the notariate made the creation of a standard map necessary. Given their role as the archivists of the resulting linguistic map, notaries were able to place their stamp on the emerging cartography. To argue for the agency and significance of late medieval notarial culture is not to diminish the subsequent role of state interest in the ensuing *furor geographicus.* It is only to say that the relationship between power and geography is indeed long-standing and that some of the linguistic and mental roots of the early modern cartographic revolution lie in the play of power and interest in the late medieval city.

CHAPTER THREE

Seigneurial Islands

I n the middle of the fourteenth century the vast majority of houses in Marseille owed to some lord or other an annual rent, called the *census*, sometimes a token amount of one penny but usually a rather more substantial figure. These rents were not like lease rents, because the proprietors (*emphiteotes*) had substantial rights over their property. By the mid-fourteenth century all urban and most rural rents in the area around Marseille were expressed in currency, although a few agricultural holdings continued to pay rent in kind. Rents came due once a year, usually on a great feast day such as Easter, St. Michael's, and especially in Marseille, the feast of the Blessed Mary on August 15 (for agricultural lands) and the feast of St. Thomas the Apostle on December 21 (for urban property). Lords also had the right to approve property transfers and demanded the payment of a *trezenum*, a thirteenth of the sale price. In Marseille, the lords, both *domini* and *dominae*, to whom these rents were paid constituted an impressively heterogeneous assemblage of institutions and private individuals, men and women, adults and children, nobles and commoners, Christians and Jews.

The bishop of Marseille, the Angevin crown, the male monastery of St. Victor, and the female monastery of St. Sauveur stand out among lords of urban property, both by reason of their substantial holdings and by the antiquity of their claims. Since 1165, when his rights had been approved by Frederic Barbarossa, the bishop of Marseille had been seigneur in his eponymous city, the *ville episcopalis*.[1] Although the temporal powers of the

---

[1] Philippe Mabilly, *Les villes de Marseille au moyen âge. Ville supérieure et ville de la prévôté* (Marseille, 1905), 4.

episcopacy had been taken over by the counts of Provence in 1257, the bishop remained the direct beneficiary of the annual rent paid by the majority of houses, gardens, and other types of property located in the upper city, particularly in the large district known as Cavalhon.[2] Some of the rents paid to the crown on houses located primarily in the lower city had once belonged to the viscounts of Marseille and had been absorbed by the counts of Provence when they reestablished their dominion over the city in the middle of the thirteenth century. The almoner and the cellarer of St. Victor owned substantial rents located in the burg of St. Catherine and elsewhere throughout the city.[3] Rent records for the monastery of St. Sauveur have not survived from the mid-fourteenth century, but to judge by other sources, the monastery owned the rents on a considerable number of houses located near the monastery in the *sixain* of St. Jean. The abbesses themselves, beginning with Maria de Tornafort prior to 1349, Laurencia Vivauda after 1349, and then Ugueta Elia after 1362, received the acknowledgments of rents owed, not delegating the task to a procurator. They defended their rights aggressively. In 1342, the Martin brothers, Bertran and Raolin, refused to pay the entry fee on a house they had acquired that was under the lordship of St. Sauveur. Hearing the news, Maria de Tornafort and "many of the nuns of the monastery," accompanied by several priests, marched to the Fishmongery where the house was located and demanded the keys. The resulting quarrel was violent enough to be described, in the transcript of the ensuing court case, as a *rix*, or mêlée.[4]

There was also, in Marseille, a host of secular lords, ranging from impressive lords like Bernat Garnier and Jacme de Galbert to lesser lords who owned a few rents on houses or buildings scattered here and there. An active market in urban rents—in which the great urban lords rarely participated—ensured the heterogeneity of the group. Nobles bought rents because to be noble was to own rents. Merchants did so in part because rents generally returned an annual figure of 5 percent on the investment, not always as profitable as merchanting, but generally safe and consistent.[5] Monastic orders, too, liked the financial security that

---

[2] In the treaty of 1257 whereby the bishop of Marseille transferred his jurisdiction to Charles of Anjou, the rents were specifically excluded. See Mabilly, *Villes*, 18.

[3] For the mid-fourteenth century, see in particular the registers of rents owed to agents of St. Victor kept by the notary Peire Gamel, ADBR 1H 1145 and 1146.

[4] ADBR 3B 808, 228r–272r, 25 Oct. 1342.

[5] Numerous sales of rents (*emptiones censualis*) can be found in notarial casebooks; in general, a rent of 50 shillings will cost 50 pounds, or twenty times more. See, for example, ADBR 381E 384, fols. 29r–v, 2 May 1337. The ratio of 1 : 20 was not fixed.

came with investing in rents. For secular men and women, simple profit was not the only motive, since the ownership of rents carried the hint of a noble way of life. Wherever status is for sale, there one will find social climbers. In mid-fourteenth-century Marseille, the list of those buying rents included not just great merchants like Peire Austria, but also a baker, two drapers, a draper's widow, two laborers, three notaries, a cobbler, a butcher, a buckler, a mason, the Jew Bondavin de Draguignan, and three other prominent Jews who had been elected to manage the Jewish almshouse (*probis viris electis super facto elemosine judeorum Massilie*).

To be a lord was to be a keeper of records, unless one's holdings were small and one's memory exceptional. The most common type of seigneurial record from the fourteenth and fifteenth centuries was a simple register or cartulary consisting of transcripts of notarial acts whereby proprietors acknowledged seigneurial rights. The notarial act itself was called a *recognitio censi* and was usually demanded by lords whenever a piece of property was transferred from one owner to another. These acts are common in extant notarial casebooks, usually found in abbreviated form immediately after the record of the conveyance itself, but sometimes independently. Particularly zealous lords could collect acknowledgments more systematically if they chose to do so, using notaries hired for the occasion to record the transaction. Following the Black Death of 1348, when a great deal of property changed hands, a number of lords conspicuously collected acknowledgments from rent-paying proprietors in order to set their books in order. In episcopal records from the late seventeenth or eighteenth centuries we find that rents and acknowledgments were collected annually, and curial officials employed a time-saving device of printing a notice to the effect that both were now due, and would the owner please present himself to the bishop's notary within a week. Blank spaces were left for the date, the name of the owner, and the location of the property.[6]

Small private lords possessing rights on a half dozen properties presumably had no difficulty knowing the extent of their holdings and whether all their annual rents had been paid. They kept copies of the acknowledgments and made a mental note, whenever property changed hands, not to harass the erstwhile owner if a rent went unpaid. The bishop, the crown, the monastery of St. Sauveur, and several other lords, in contrast, owned dozens or hundreds of rents in the city alone. The

[6] ADBR 5G 129 (liasse dated 16 May 1653); see also 5G 108 for slips dated 1679.

challenge for these lords was to create records that permitted systematic crosschecking: Did the person who now owns this house located in this area pay this year's rent? Proprietors may show up conscientiously on the feast of St. Thomas the Apostle to acknowledge the servility of their property and pay their rent, but they do not organize themselves carefully in alphabetical order or neatly by street or block or neighborhood as they wait in line outside the gates of episcopal, royal, or monastic buildings. The resulting record of receipts or acknowledgments, if kept at all, will be entirely random unless the acts are ordered according to some rubric as they come in. Record-keeping was complicated by the rapid turnover of proprietors; any systematic record of known proprietors will inevitably become out-of-date with the passage of time.

Lords developed several methods for overcoming this record-keeping challenge. The solutions took into account that whereas proprietors were changeable, the properties themselves were permanent. One method, used in Marseille and elsewhere in medieval Europe, was to create an index of the random record of acknowledgments by street or island or any other location, so that any interested official could easily find all the acknowledgments for houses located, say, on the island of Durant Barbier, and see if any were missing. A second and more labor intensive method involved organizing the acknowledgements by location as they were received, or perhaps by recopying them in order. Both depended on a third and more elaborate type of record that in Marseille was called a *levadou*. This was a master record of all houses or properties grouped according to some geographic nomenclature. From this record, an official could rapidly assess the totality of the lord's current holdings and easily determine who had and had not paid the required rent. Given the turnover of proprietors, officials took to writing, underneath each registered house, the genealogy of successive owners, indicating the date of the transfer. One episcopal register in French from the sixteenth and seventeenth centuries even used the expression "genealogie" to describe the nature of this record.[7]

Perhaps the best example of seigneurial record-keeping from mid-fourteenth-century Marseille, a form of double-entry bookkeeping, can be found in two registers used by the merchant Bernat Garnier. One register is a cartulary, in Latin, in which we find a record of all acknowledgments collected whenever property under Bernat's dominion changed hands.[8] The acts chronologically follow one another, beginning

---

[7] ADBR 5G 127.
[8] ADBR 4HD B1quater.

in 1315 and running up to 1345. An index at the front—incomplete, as it happens—groups the houses by street or area; it tells the reader, for example, that transactions relating to all properties located "behind St. Marthe" can be found on folios 6v, 8v, 9r, 15r, 16r, 18v, 51r, 84r, and 104r.

The second is a *levadou*, kept in Provençal; here, all Bernat's urban and rural properties were organized according to their location, using the same rubrics that appear in the index to the cartulary.[9] Under each property we find a genealogy of the owners. Below is a typical entry, identifying an *ostal* in the quarter known as Rocabarbola.

En rocabalbola Jacme Bonier e sa molher fan de sensa pro una ostal lo cal sensa pagas en la festa de san Tomes apostol.	In Rocarbarbola, Jacme Bonier and his wife paid the rent for a house; the said rent is due at the feast of St. Thomas the Apostle.
Vendet a Guilhem Robol en l'an MCCCXXVII e lo men de Otambre fec la carta en Jacme Lautaut.	Sold to Guilhem Robol in the year 1327 in the month of October; Sir Jacme Lautaut [the notary] made the act.
Vendet a Huguo Depagas en l'an MCCCXXVIII en lo mes d'Aust fec la carta en Jacme Lautaut.[10]	Sold to Huguo Depagas in the year 1328 in the month of August; Sir Jacme Lautaut made the act.

Thus, whenever a piece of property changed hands, Bernat had the notary copy the act acknowledging his lordship into his cartulary. Then someone, probably Bernat himself, copied the gist of the act into the appropriate place in the *levadou*.

The fundamental organizing principle of the cartulary was chronology. In contrast to this, the fundamental organizing principle of the *levadou* was topography. Simply to index the records required the development of common cartographic categories that once fixed in record-keeping practices, acquired permanence over time. When great urban lords or their officials leafed through old records or mentally surveyed present-day dominions, they saw familiar houses clustered together in well-known areas and inhabited by ephemeral genealogies of humanity. As a result,

---

[9] ADBR 4HD B1ter. The *levadou* also includes agricultural lands, and these were similarly organized by location.
[10] Ibid., fol. 40r.

the records left by the lords or their scribes are suffused with a linguistic cartography.

The surviving registers concerning urban property kept by or for private lords in the fourteenth century reveal a cartography that is, for the most part, little different from the one used by the notaries: streets predominate, but other nomenclatures are used, notably nearby landmarks. This is to be expected. Many private lords, like Audoart Alaman, hired public notaries to collect both rents and the written acknowledgments of lordly rights from householders; in these cases, notaries used their own emerging cartography in writing site clauses.[11] More substantial (and literate) private lords like Marie de Jerusalem, her son Raymon del Olm, and Bernat Garnier compiled their own registers in their own language, epitomizing the relevant notarized acknowledgments.[12] It made sense to borrow the cartographic categories of the Latin, notarial original.

Unlike these private lords, the two great institutional lords who left systematic rent records, the bishop of Marseille and the Angevin crown, possessed their own curias. These curias employed private notaries who, in speaking of urban property, used a distinctive cartographic language. This was the language of *insulae*, islands or city blocks, also used by officials of the city council to organize the records of direct taxes. The insular template accomplished for these lords or their curial officials what the street template accomplished for notaries: it identified the location of a piece of urban property. Like the street template, the insular template used abutment clauses to specify the location of the property on the block. In crown records, for example, one Meralda de Jerusalem acknowledged owning a house on the island of Laugier de Soliers. It was a thin block, and the abutment clause lets us know that her house had doors opening on both sides, and touched the houses of Uguo dal Temple and Antoni de Ricasnovas on one side and Johan Arnes on the other.[13] Given the use of the abutment clause, the insular template was just as precise as the street template in locating property on the mental map.

The use of the insular template owes much to its antique origin, for islands were an important component of the linguistic and administrative

---

[11] ADBR 355E 3, fols. 7r–v. Four acknowledgments of rent owed to Audoart appear one after the other in a casebook of the notary Jacme Aycart.

[12] The register of Marie de Jerusalem and her son Raymon del Olm can be found in ADBR 1HD H3.

[13] ADBR B 831, fol. 5r.

cartography of ancient Rome.[14] The template probably survived in the bureaucratic language of western Christendom, heir to Rome in so many ways. The use of islands fits well into the intellectual hierarchy of Christian administrative topography; after all, the custom of parceling space into discrete territories like bishoprics and parishes was a practice of long standing.[15] The practice, moreover, extended outside the realms of ecclesiastical administration, for the language was a convenient one, and as a result islands occur frequently in medieval records that locate urban property sites.[16]

In Marseille, we find islands in the earliest episcopal and royal rent registers from the late thirteenth century, and they dominate the records of these two great lords throughout the fourteenth and into the fifteenth centuries. Episcopal and royal curias, moreover, were much more consistent in using their favored template than were public notaries in using theirs. In 177 site clauses from an episcopal register compiled between 1343 and 1347, for example, 157 or 89 percent were based on the insular template. The curial notary who compiled the next record between 1353 and 1359 did better, using islands in 264 of 281 site clauses, or 94 percent of the time.[17] The notaries, who used streets or open spaces in only 58.3 percent of their site clauses, come off as less consistent in their cartography. The *clavaire*, the royal official in charge of collecting crown revenues, was similarly committed to islands, although a few areas on his linguistic map were listed by street, not by island, for reasons that are not clear.[18] These seigneurial officials, in other words, were also professional cartographers, using a template much more consistently than did the public notaries.

It is all the more interesting, as we will see below, that both episcopal and royal officials had for the most part abandoned the language of islands in their records by the early sixteenth century. Although records continued to be organized geographically, officials had switched over to the street template, which, in notarial usage, had triumphed

[14] O. A. W. Dilke, *Greek and Roman Maps* (Ithaca, 1985), 88; Claude Nicolet, *Space, Geography, and Politics in the Early Roman Empire*, trans. Hélène Leclerc (Ann Arbor, 1991), 135.

[15] Robert Bartlett, *The Making of Europe: Conquest, Colonization and Cultural Change, 950–1350* (Princeton, 1993), 5–23.

[16] See, for example, Henri Broise, "Les maisons d'habitation à Rome aux XVᵉ et XVIᵉ siècles," in *D'une ville à l'autre: structures matérielles et organisation de l'espace dans les villes européennes (XIIIᵉ–XVIᵉ siècle)*, ed. Jean-Claude Maire Vigueur (Rome, 1989), 609–29. We also find islands in Languedoc; see André Gouron, *La réglementation des métiers en Languedoc au moyen âge* (Geneva, 1958), 127, n. 65.

[17] ADBR 5G 112 and 5G 114.

[18] See, for example, ADBR B 1940, B 1941, and B 1942, redacted between 1331 and 1359.

over island, vicinity, and landmark several decades earlier. The transformation had been underway in seigneurial circles since the early to mid-fifteenth century. The changing habits of successive generations of seigneurial officials thus show how they collectively were influenced by the process of cartographic universalization that first arose in public notarial circles.

The Cartography of the Episcopal Curia

Most episcopal rents were derived from properties located on islands situated to the west of the street of the Upper Grain Market in the unofficial district of Cavalhon. The islands of the upper city found in this area took their names, for the most part, from prominent inhabitants: Uguo Pauli, Guilhem Sard, Fulco Sardine, Berengier Repelin, and so on. In the mid-fourteenth century a number of the eponymous individuals were still living, and the names of several islands were clearly being passed down from father to son, such as the island named after Uguo de Scala that later took up the name of his son, Guilhem. A few islands were named for the grain market in the upper city; this included the island of the Upper Grain Market (*insula Annonarie Superioris*), the island behind the Grain Market (*insula retro Annonaria*), and the uneuphonius island on the Other Side of the Grain Market (*insula ab alia parte Annonarie*). Several bore the names of nearby churches or gates, such as the island of the Swords, named after the hospital and church of St. Jacques of the Swords. One was named after the Jewry of the upper city. Although the community itself was moribund by the 1340s, the name continued to be used in episcopal records throughout the fourteenth century. Several island names were also, on a few occasions, the names of streets—the street of the Upper Grain Market, the street of St. Jacques of the Swords, the street of Pons Broquier, the street of Cavalhon, and two or three others.

Site clauses typically described a property site located in such-and-such an island and adjoining certain properties. Thus, in 1353, the laborer Durant Augier acknowledged owing rent to the bishop for a house located in the island of Jauceran the Saddler (*insula Jaucerani Manescalli*), touching a house belonging to Guilhem Adhemar, a water well, a house belonging to Peire Chabas, and an unnamed public street.[19] From the abutment clauses we can tell that most blocks were broad. Houses did not extend through the block, and instead adjoined another house at the back. Site

---

[19] ADBR 5G 114, fols. 6v–7r.

clauses occasionally used a nesting style, as when Esteve Broquier acknowledged a rent on a house located "in Cavalhon in the island of Pons Broquier."[20] This can be found only in several dozen cases involving houses
located in Cavalhon, never in the other major upper-city district of
Rocabarbola. Nor was it used consistently within the rough limits of
Cavalhon.

Notaries who worked in the episcopal curia were conscious of the need
to organize acknowledgments of episcopal lordship both topographically
and chronologically. A *levadou de cens* from the year 1353 provides a listing
of all vineyards and fields under episcopal lordship that are organized by
rural location, such as "Ad fontem del Leon," "Altor de Bonafos," "Balma
Maynart," and "de Canneto."[21] In terms of urban property, a register comparable to a *levadou* is extant from the year 1343. This record numbers
each island from one to twenty-six and then, under the rubrics that follow,
provides the names of the proprietors and the rent each one owes.[22] An
index similar to a *levadou* was copied out at the beginning of a record of
acknowledgments of rent for property both urban and rural compiled
between 1343 and 1347.[23] In this register, the unnamed notary identified
forty-one islands in the upper city in which the bishop owned rents,
including two islands located in the suburbs just outside the walls. Under
the rubric for each island, he listed the names of all the current property
holders.

Despite the fairly systematic record-keeping practices of the episcopal
curia, there were a few inconsistencies in the linguistic map that lay
behind the documents. For example, the list of forty-one islands at the
beginning of the 1343–1347 register does not quite correspond to the
islands listed in the site clauses of the ensuing acknowledgments: some
islands in the list do not show up in site clauses, and the site clauses
include several references to streets and landmarks. As it happens, a
similar register was compiled in the wake of the plague, between 1353
and 1359. Perhaps because of the recent catastrophe, this one was more
thorough—among other things, the number of properties has risen from
177 to 281, indicating either that a number of properties had been missed
in the previous compilation or had been acquired by the episcopal curia.
All the islands listed between 1343 and 1347 show up in the site clauses

[20] Ibid., fols. 8v–9v.
[21] ADBR 5G 115.
[22] ADBR 5G 113. The enumeration of the islands is on fol. 5r; similar lists from 1344 and
1345, also numbered, are repeated on fol. 68r and fol. 89r.
[23] ADBR 5G 112.

in between 1353 and 1359, and few site clauses did not use the insular template.

Properties were not distributed evenly among the forty-one islands. Most were located in some twenty islands in the region of Cavalhon where episcopal holdings were especially dense. These islands were the most conscientiously mapped according to the insular template. For the time being, the map was durable: most of these islands, the names unchanged, can be found in other registers produced by the episcopal curia in 1391 and between 1423 and 1425.[24] The islands or streets with only one or two houses in the 1353–1359 register, in turn, have dropped out of the picture. The rents to these houses may have been sold in the intervening years.

Islands are curious topographical entities. With streets, it is relatively easy to reconstruct a medieval urban map. Every now and again one will come across site and abutment clauses describing houses located at intersections, and as these accumulate one is able to rebuild the entire tracery of streets. Islands, as the name implies, float independently of each other, and in theory it is never possible to discover how they stand in relation to one another. On rare occasions, curial notaries using the island template made reference to facing islands or landmarks. The island of Guilhem Sard, for example, faced the episcopal palace, and the expression "the island of Guilhem Sard facing the episcopal palace" (*insula Guillelmi Sardi ante domum episcopalem*) is relatively common in episcopal records. But in general, to find out where the islands of the upper city were located one must first establish the tracery of streets and then use abutment clauses in episcopal records to establish where the islands stood in relation to the streets.

Even this is difficult. One of the most curious features of the cartographic habits of the episcopal curia is that curial notaries rarely identified streets even in abutment clauses. Almost invariably, where an abutment clause describes a street, the street is an anonymous public street (*carreria publica*) located in front of or alongside the house in question, or perhaps an anonymous alley (*transversia*) to the side or rear. Using site and abutment clauses in mid-fourteenth-century episcopal records, we can identify only eight named streets or alleys in the area of Cavalhon where most episcopal rental properties were located. To judge by the map of the city, this area should have had several times that number of streets. The addresses offered by some of the proprietors provide a few more, but otherwise, from the point of view of the toponymy of streets, Cavalhon

[24] See ADBR 5G 119 (1391) and 5G 126 (1423–25).

was largely unmapped. This was the product of the custom of the episcopal curia to take islands as the fundamental unit of cartographic awareness.

The unmapping of Cavalhon, from the perspective of the modern historian, is one of the most bizarre of all cartographic phenomena in fourteenth-century Marseille. The map of the lower city is fairly well known, thanks to generations of painstaking cartographic research. J. A. B. Mortreuil, Octave Teissier, and Bruno Roberty may have disagreed on street names in several places, but for the most part there is common agreement in their maps on the architectonic form of the lower city—the layout of streets, squares, and blocks and the location of buildings, gates, and walls. This plan was derived from modern cadastral maps, and there seems to have been little change in the spatial plan in the intervening centuries. In the upper city, all is confusion. On both Teissier's and Roberty's maps we find only eight or ten streets. The plans themselves disagree not just on the names assigned to given streets but also on the spatial plan of the entire area. Roberty's map shows that as he puzzled his way through the documents, he added streets into his original plan in an effort to make sense of the topographic information given in his sources. He also left open spaces in the lines marking streets where he knew there to be intersections, but his information did not allow him either to fill in a street name or to guess where the intersecting street led. The verbal map drawn by Philippe Mabilly is little better, listing only a few streets and confusing islands with streets. Yet it is unlikely that the spatial plan of the upper city changed much more than that of the lower city, despite important construction projects in the seventeenth century. The confusion arises from the fact that the upper city was, from the perspective of streets, largely unmapped, or at least was mapped in a wholly unfamiliar way.

The absence of street names sometimes led public notaries to invent odd expressions to identify streets. We find entities named "the street that goes to the oven," "the street that goes to the said church [of Sedis]," "the alley that goes to the sea," and "under the Gentleman's Oven, in the street that leads to the bishop."[25] This kind of circumlocution, as Leguay indicates, is not uncommon in the practice of thirteenth- and fourteenth-century public notaries and scribes. As far as Marseille is concerned, it was just more common for properties located in Cavalhon.[26] The problem became especially acute in the sixteenth and seventeenth centuries, by

---

[25] ADBR 351E 645, fols. 18v–19r, date unreadable; 351E 642, fols. 55r–56r, 20 Apr. 1358; 355E 11, fols. 2r–v, 31 Dec. 1360; 381E 76, fols. 11v–12r, 8 Apr. 1347.
[26] Jean-Pierre Leguay, *La rue au moyen âge* (Rennes, 1984), 93.

which time episcopal notaries had abandoned islands in favor of streets. In the course of drawing up site clauses in both French and Latin, they realized, as if for the first time, that many of their streets had no names. We find houses located on "a street leading from St. Antoine to St. Cannat," "a street under St. Jacques," "an alley descending from St. Paul," "a street descending from the Carmelites to St. Clare," "a street leading from Colla to the episcopal compound," "a street rising up from the street of the Guiberts toward the Carmelites," "a street leading from the church of St. Jacques of the Swords to the Porte d'Aix," and, most remarkably, "a street named [blank space] descending toward Cavalhon" (*en la rue dicta* [blank space] *descandans vers Cavailhon*).[27]

The lack of street names, let alone names of vicinities, created an interesting situation in which medieval residents of the upper city tended to identify themselves with the large district of Cavalhon; rarely did they draw a more fine-grain map of their residential area. Of the 423 individual men and women listed in the episcopal register of 1353–1359, a total of 141 simply named Cavalhon as their place of residence. An additional 122 listed a street as their address, including many streets located in the lower city; only seven of these named a street located in Cavalhon. The insular template was almost never used in the addresses found in any notarial act, rent register, or similar record from late medieval Marseille. In the episcopal register of 1353–1359, only four people actually used the insular template in giving their addresses, and one of these islands was located in the lower city.

It is not clear why the residents of Cavalhon did not subdivide the area into its constituent streets. We know, from the abutment clauses given in all episcopal registers, that there were plenty of streets running through it. Only twelve streets are named in the 1343–1347 register and of these only five, the street of Cavalhon, the street of the Old Market, the street of the Well, the street of Ribot, and the street of St. Antoine, were located in the area largely under the cartographic control of the episcopal curia. In the 1353–1359 register there are also five streets in Cavalhon, the street of Isnart Beroart, the street of Guilhem Naulon, the street of the Well, the street above St. Antoine, and the street of Francigena.[28] It is possible that the laborers who lived in Cavalhon did name their own streets but that these names were not recognized by notaries of the episcopal

---

[27] These examples are drawn from ADBR 5G 135 (compiled in the fifteenth and sixteenth centuries), 5G 142 (compiled between 1512 and 1557), and 5G 157 (compiled 1668–1671). The last example is 5G 142, fol. 116r.
[28] This list does not include the long street of the Upper Grain Market. The bishop controlled rents along part of it but not its whole length.

curia, although this is unlikely since the names do not show up elsewhere. It is also possible that the hilly terrain of the upper city prevented the formation of the relatively straight public ways that were called streets in the lower city.

The most likely explanation, however, lies in the lack of social differentiation in Cavalhon. The prosopographical index reveals that, in general, around 65 percent of the several thousand residents of the upper city were laborers. The 1353–1359 episcopal register itself confirms the figure: of seventy-one known residents of Cavalhon whose profession is indicated in the register, forty-six or 65 percent were laborers, with their residences distributed widely throughout the district.[29] The dominance of laborers throughout Cavalhon probably hindered the formation of the knots of artisanal or vicinal sociability that so profoundly influenced toponymy in the lower city.

When added to the failure of the episcopal curia to develop a cartography acceptable to the laborers of the district, we have all we need to explain the unmapping of Cavalhon. As we saw in chapter two, the imaginary cartography of an area was less systematic whenever public notaries, the nodal agents of cartographic conversations, did not use their authority to establish a consistent cartography. This was the case with the suburbs, areas in which relatively fewer notaries lived and walked, areas that therefore fell somewhat outside the sphere of notarial cartographic competence. In the upper city, by contrast, we find a very consistent cartography established by long-standing curial practice. Moreover, there were two mechanisms for conveying this information to proprietors and citizens. First, were the yearly occasions on which proprietors queued up to pay rents and to acknowledge the servility of their properties. Second, episcopal officials encouraged notaries and clients to transact property conveyances in the physical space of the episcopal compound, occasions on which the episcopal map must have been used and therefore publicized.

As mentioned earlier, however, a striking feature of notarized property conveyances in Cavalhon is that public notaries were much less likely to use the language of islands when they wrote up their acts in neutral spaces. One such conveyance, transacted in one of the law courts, listed the property site as "the street of St. Jacques of the Swords" (*carreria St. Jacobi de Spatula*), which was also the name of an episcopal island, while

---

[29] ADBR 5G 114. The only other well-represented occupational group was the fishermen, who totaled eleven, or 15 percent of the total. No other occupational group was represented by more than two members.

another conveyance, transacted in the house of a buyer, a nobleman called Johan Naulon, listed the site as "the street of Paul Naulon," who, we can guess, was a relative of Johan.[30] Hence, the space of the transaction has a bearing on cartographic usage: episcopal islands were more or less confined to curial spaces. As with public notaries, so with the general population, and, in summarizing the evidence above, we can now say confidently that episcopal islands simply did not seep into everyday cartographic language. The reasons for this resistance are unclear, though it may have been because islands were, for the laborers and fishermen of Cavalhon, politically unsympathetic cartographic entities, a constant reminder of episcopal seigneurial power. But since the episcopal curia dominated the nodes of cartographic conversations concerning Cavalhon, and required that public notaries transact property conveyances in episcopal spaces, a consistent cartography based on streets or even vicinities never developed in the fourteenth century. Cavalhon itself became a kind of super-vicinity, and, for many or most residents of the region, became their address by default.

THE CARTOGRAPHY OF THE ROYAL CURIA

Although the Neapolitan kings and queens were relative newcomers to the county of Provence and its revenues, they collected rents at one time paid to the viscounts of Marseille on houses located primarily in the lower city, rents that were probably as ancient in origin as those collected by the bishop. Other rents were more recent, having been seized from the estates of the merchants, such as the infamous Johan de Manduel, who had rebelled against Charles of Anjou in 1263. Technically, all rents were paid to the count of Provence or his representatives, and therefore show up in the financial records of the seneschal of Provence. Like the rent registers of the episcopal curia, those maintained by the royal *clavaire*, the chief financial officer of the Angevin crown, organized the houses owing rent primarily according to islands. We find islands in the earliest royal records from the late thirteenth century, and insular usage persisted into the fifteenth.

All royal records are organized according to geography, and unlike the records of the episcopal curia, royal records include only the essentials of the acknowledgment, not the entire act. The number of islands in royal

[30] See ADBR 351E 642, fols. 16v–18r, 24 Sept. 1351; ADBR 381E 82, fols. 185v–186r, 19 Nov. 1359.

records varies considerably from one register to the next, and the reasons for this are not clear. The three registers compiled between 1331 and 1359 include around 320 houses grouped into the same nineteen islands.[31] In contrast, a register from 1377 lists 487 houses grouped into fifty-one islands or other sites, and an earlier register, from 1301, includes thirty named islands and several other sites.[32] The islands, to the extent that we can identify their locations, were scattered all over the lower and upper cities, with concentrations along the port and in the southeastern part of the city.

Like episcopal notaries, the notaries who worked for the *clavaire* were fairly consistent in grouping houses by island. In some cases, a few miscellaneous houses acquired a different sort of label. In the 1301 register, for example, one heading was given as "Concerning the rents on houses that used to belong to the late Johan de Manduel located in the street of the Almoner."[33] These and several other houses had come recently into the possession of the crown and had not acquired insular names. In later records, some of these new houses were reorganized into islands. In 1301, we learn about "the rents of Tholoneum" (*de censibus Tholoney*) that were once paid to the rebel Anselm Feri and that had fallen into the hands of the crown.[34] The houses were located "in a street that is now called the street of Guilhem de Sant Gilles." In later records, these same houses always show up as being located on the island of Guilhem de Sant Gilles or that of his son, Antoni.[35] Hence, we see an interesting shift in cartographic language from the vicinity known as Tholoneum to the island of the Sant Gilles family, with the street being an intermediate phase. The notaries of the royal curia, like other agents, were in the process of creating a standard cartographic language, even if the trajectory—from street to island—differed from the notarial trajectory.

As this last example indicates, a few houses located next to one another on a street were capable of being identified as an island. In such cases, there is very little difference between island and street; hence, the *carreria Guillelmi de Sancto Egidio* was entirely interchangeable with the *insula Guillelmi de Sancto Egidio*. In other cases, we can tell from abutment clauses that the crown received rents from all or most of the houses on a given block. An island named after Raymon de Arbore was defined in one record as being located within the streets of the Almoner, of Castilhon,

---

[31] ADBR B 1940–1942.
[32] ADBR B 831 (dated 1377) and B 1936 (dated 1301).
[33] ADBR B 1936.
[34] Ibid.
[35] For example, ADBR B 1940 (1331), fol. 176v.

of the Curriers, and of d'an Galli; clearly, this was a block and not a fancy name for a street frontage. Eight houses located within this block are listed in order. The abutment clauses show that the first abutted the second, the second the third, and so on. The first and eighth houses, in turn, also abutted each other, showing that we have figuratively just walked around the block.[36]

Some streets were obviously more significant than others, and in some cases islands were separated only by the thinnest of alleys. As a result islands, in some cases, were named after the major street on which they were located. The island of the Changers, for example, fronted on the street of the Changers. The street of the Carpentery, which ran down a thin intramural space, created cartographic problems, since notaries of the royal curia were not sure whether to define the area as the two islands of the Carpentery, one of which touched the old wall (*insula Fustarie barrii veteris*) and the other the new wall (*insula Fustarie barrii novis*), or whether to call the whole area the street of the Carpentery (*carreria Fustarie a parte barrii veteris, carreria Fustarie a parte barrii novis*). In most cases, however, islands did not share names with the streets that surrounded them. As in episcopal records, the majority bore names of significant proprietors. These changed regularly; thus, we hear of the "island of Gautelm Malet that is now Antoni Malet his son's" and "the island that used to be called Peire de Berre's and Peire Blancart's and is now Johan de Gemenos'." In the fourteenth and early fifteenth century there apparently was a close relationship between an island and its eponymous inhabitant: one Antoni Bonfilh was described as "holding the principal place" on the island named for him.[37] Islands were sometimes identified as being located "in" a street; thus, the island of Carle Athos was located in the Greater street of the Jerusalem and the island of Giraut Lort in the street of the Tannery.

The coupling of the use of islands with patrons or family lineages may be the key to explaining their popularity among the two great lords in Marseille, as well as elsewhere in Europe. When the domestic architecture allowed, the implied sociability of the island focused inward toward a courtyard or shared living space that was dominated, as in ancient Rome, medieval Genoa, and certain modern cities in Islamic areas, by a kin-group or a patron. Thus it turned a blank wall to the exterior.[38] A

---

[36] ADBR B 831, fols. 64r–65r.

[37] ADBR B 1177, fol. 34r. The phrase is *cuius insule ipse Antonius tenet principium*.

[38] Jacques Heers, *Espaces publics, espaces privés dans la ville. Le liber terminorum de Bologne (1294)* (Paris, 1984), 33–45; Janet L. Abu-Lughod, "The Islamic City—Historic Myth, Islamic Essence, and Contemporary Relevance," *International Journal of Middle East Studies* 19 (1987): 167.

similar sociology may have operated even in a place like Marseille where
courtyards were uncommon. As the work of Henri Broise and Jean-
Claude Maire Vigueur suggests, the idea of an island requires the street
itself to become merely a space between compounds and not a thoroughfare, which is to say, an open space permitting urban circulation.[39]
Because medieval lords or family lineages exerted control most effectively
through chains of patronage, the template of the island fit their political
goals.

As in the case of episcopal islands, royal islands almost never found
their way into everyday usage. In their casebooks, the public notaries
never used the language of islands in drawing up site clauses involving
lower city property, in part, undoubtedly, because they were never
encouraged to transact property conveyances in Angevin spaces. By the
same token, in all the records from the mid-fourteenth century, I can find
only one person who gave his address by means of a royal island. This was
the butcher Guilhem Symon. In 1353, while acknowledging episcopal
lordship of several possessions, Guilhem gave his address to an episcopal
notary as the "island of Sant Gilles" (*insula de Sancto Egidio*), a royal island
located in the lower city.[40] In a royal register from 1351, two years earlier,
we find Guilhem acknowledging a house located in this same island,
undoubtedly the house he considered his home. Possibly the memory of
this recent exchange with the royal notary prompted him to give the
island as his address to the episcopal notary in 1353.[41]

Yet this was an exceptional case. In both the upper and lower cities we
find an aversion on the part of the general population to using the
seigneurial language of islands. In the upper city, evidently, no alternative cartographic language had ever been developed, at least by the fourteenth century. This was not the case in the commercially vibrant lower
city where an active group of public notaries had been supervising the
emergence of a regular map for some time and where powerful craft
organizations had long since stamped their own cartographic style on the
city map. As a result, several alternative cartographic languages were available. Nowhere is this clearer than in an exceptional royal record from
January 1318.[42] Unlike other records of the seigneurial rights in property
of the Angevin crown, almost all of which were neatly organized by island,

[39] See Henri Broise and Jean-Claude Maire Vigueur, "Strutture famigliari, spazio domestico
e architettura civile a Roma alla fine del medioevo," in *Storia dell'arte italiana* (Turin, 1981),
99–160.
[40] ADBR 5G 114, fol. 56r.
[41] ADBR B 1941, fol. 11r.
[42] ADBR B 819.

this one was a fragmentary and hasty record of acknowledgments of seigneurial rights as they were received. The record lists the location of the property but the notary or scribe made no effort to translate these sites into the language of the royal curia. There is only one island listed in this record, the island of Peire de Berre. Instead of islands, what we are reading is evidently the cartographic language of the average Provençal speaker: seventy-seven of these people named vicinities, forty-eight gave streets, and people mentioned ten other types of sites. The vicinities do not correspond to the islands usually found in crown records. For example, there are twenty houses located in the vicinity known as the Tannery (*Blancaria*). In other crown records this vicinity is invariably broken up into its constituent islands, such as the island of Antoni Bonfilh, the island of Guilhem or Giraut Lort, the island of Peire de Berra or Peire Blancart, the island of Johan de Cuges, and several others. Another vicinity used on four occasions in the 1318 register, the Jewish fountain, does not appear as an island in other royal registers. The names of the streets more frequently match the names of islands, such as the street of the Johans, the street of Johan de Serviers, and the street of the Sant Gilles, all three of which were also the names of islands. Apparently, ordinary Provençal speakers were oblivious to islands.

The Cartography of the City Council

The city council of Marseille was not a property lord. All the same, council officials occasionally used the language of islands in fiscal records, which is why I include council cartography in this chapter. The most conspicuous cartographic language of the city council was the language of the *sixains*. The administrative *sixain* in which a citizen resided mattered in several ways. Eligibility for council membership, for example, was determined by *sixain*, as was liability for infrastructural repairs.[43] In the calamitous 1350s we find that men were being drafted for the military defense of the city according to the *sixain* in which they lived, and so the *sixains*, like the *gonfalons* of Florence or the parishes of Venice, could and did have a military function as well.[44]

---

[43] Citizens of a *sixain*, for example, could be assessed for the cleaning of a fountain in their *sixain*. See AM BB 21, fol. 137.

[44] For Florence, see the general remarks of D. V. and F. W. Kent, *Neighbours and Neighbourhood in Renaissance Florence: The District of the Red Lion in the Fifteenth Century* (Locust Valley, N.Y., 1982), 4; for Venice, see Dennis Romano, *Patricians and Popolani: The Social Foundations of the Venetian Renaissance State* (Baltimore, 1987), 19.

Most important, the *sixain* was the primary unit for purposes of direct taxation, both for the general taille (Prov. *talha*; Lat. *tallia*) and for a capitation.[45] A complete register of one of the general tailles has survived for the fiscal year 1360–1361; in it, the two thousand assessable individuals were all identified by means of their *sixain* of residence.[46] City council minutes reveal that six or possibly twelve officials known as *seyzenerii* were assigned the responsibility of supervising the collection of the taille in each of the six *sixains* in the lower city.[47] As soon as the city council had fixed the amount to be collected in a given taille— the amount was not constant from one taille to the next, and could go up considerably during periods of warfare—the *seyzenerius* made his assessment (*fecit suam talliam*) and then turned the process over to officials known as *illerii* who would actually make the collection door-to-door. *Illerius*, of course, means "islander," that is to say, someone in charge of collecting payments from inhabitants of a given island. The association is made clear in the partial register of another general taille collected in 1384–1385.[48] The treasurer registered each payment made by the collector by means of the following formula: "On this date, so-and-so, the islander of his island (*illerius de sua insula*), paid to the treasury so many florins."

These two tax registers underscore the fundamental point that the council had developed a sophisticated and consistent fiscal cartography that, in turn, complemented the general political cartography of *sixains*. Since the council used the cartography of islands, it clearly borrowed from episcopal-royal cartographic usage. However, perhaps the most striking feature of council cartography was its near total failure to influence anyone. Notaries almost never used the language of *sixains* in describing site clauses. Individual men and women, by the same token, almost never used *sixains* in the addresses they provided in identity clauses. The islands used by the city council were never even named, and therefore naturally do not figure in other cartographic styles. Like episcopal and royal islands, council islands were closely identified with a fiscal regime and were rarely understood as a natural cartography.

The important point in all this is that the council, perhaps the most

---

[45] Alain Droguet, *Administration financière et système fiscal à Marseille dans la seconde moitié du XIV* siècle (Aix-en-Provence, 1983), 33–36.

[46] For this register, see AM EE 55 A.

[47] See AM BB 22, fol. 63r.

[48] AM CC 175.

obvious form of government in fourteenth-century Marseille, was not generating a cartography that would become an accepted or conventional pattern of property and personal identity. Island and *sixain* were understood to be addresses relevant for fiscal, political, and administrative purposes, but, as in ancient Rome, such addresses did not play any role in identity clauses. This is understandable. Few people can have enjoyed paying taxes in fourteenth-century Marseille. For the vast majority of male residents of the *sixains*, let alone all women and Jews, there was never any question of joining the city council. The most significant rights enjoyed by the citizens of Marseille were those accorded to the whole citizenry. As Mireille Zarb suggests, Marseille's privileges amount to a kind of patrimony, shared and defended by all.[49] One's *sixain* of residence was largely irrelevant to this larger identity.

Marseille's experience in this regard roughly fits in with the developing norms of the *ius commune*. As Julius Kirshner has argued, the jurist Bartolus de Sassoferrato (d. 1357), concerned to promote the uniformity of citizenship throughout a city, "rejected all claims of sectional citizenship within the *civitas.*"[50] Such opinions, of course, did not preclude the development of emotional bonds to a given section. In Florence, for example, the relationship between *de facto* political rights and quarter of residence was somewhat tighter and quarter identity correspondingly stronger, owing to the exercise of patronage.[51] In the absence of systematic studies of identity clauses in Florence and other European cities and towns, however, it is difficult to draw hard and fast conclusions about Marseille's place in larger trends.

To sum up, the cartographies generated in Marseille by the episcopal and royal curias and by council officials, although sophisticated, never spread beyond the narrow domain of fiscality and politics. This is particularly remarkable, given that ordinary citizens had regular contact with this cartography. Over time, as notarial cartographic hegemony increasingly made streets the dominant element in the urban cartographic imagination of Marseille, the fiscal cartography of the island would disappear entirely.

---

[49] Mireille Zarb, *Histoire d'une autonomie communale: les privilèges de la ville de Marseille du X*^e *siècle à la Révolution* (Paris, 1961), esp. 32–40.

[50] Julius Kirshner, "*Civitas Sibi Faciat Civem:* Bartolus of Sassoferrato's Doctrine on the Making of a Citizen," *Speculum* 48 (1973): 705–7.

[51] See, for example, Dale Kent, *The Rise of the Medici: Faction in Florence, 1426–1434* (Oxford, 1978); F. W. Kent, *Household and Lineage in Renaissance Florence: The Family Life of the Capponi, Ginori and Rucellai* (Princeton, 1977).

The Decline of the Insular Template

In and around 1350, the vast majority of rents under episcopal and royal lordship were organized in the pertinent records by means of islands. By 1500, usage of the template was vanishing. To judge by the surviving evidence provided by royal records, this change in cartographic imagination proceeded in two stages. From the late fourteenth century, registers of royal accounts and royal rights, although still fundamentally organized by island, began to include abutment clauses that named adjoining streets. Usage of street names in abutment clauses became more and more common. The habit began first in registers of royal rights and later spread to registers of accounts. Between 1450 and 1470, royal records were reorganized according to streets. In episcopal records, the transition took place over a similar time period, although in isolated registers the language of islands can be found as late as the seventeenth century. Both transformations correspond roughly in time to the transformation we have seen in the case of notarial practice, as open spaces progressively became known as streets, plazas, or alleys over the fifteenth century, and lost, in the process, not only their Provençal status as vicinities or landmarks but also their status as islands in seigneurial cartography.

In royal records, houses owing rents to the crown can be traced as far back as 1264 when the earliest record of royal property in Marseille, entirely in Provençal, was compiled. Entitled "Revenues of the City of Marseille," it includes lists of rents owed for 538 houses and other urban properties.[52] The majority of these houses were organized under the rubrics of fifty-eight islands. The notary simply named the appropriate island, such as the *isla d'Augier de la Mar* or the *islla de Ugo Andrieu*, listed the houses found on the island, and moved on to the next. The only exceptions to this insular cartography were isolated houses. Among these were *las possesiontz dels faiditz e dels condepnatz* ("the possessions of the traitors and the condemned men") as the register calls them, houses the crown had recently seized from the estates of the rebels of 1263, Aubert de la Vainna, Uguo Vivaut, and Johan de Manduel. Of the four houses mentioned, one was located in the street of Sir Giraud Amalric (*en la careria d'en Giraut Amalric*), a second in the Fruitery (*en la Frucharia*)—linguistically not a street, of course, but equally not an island—a third in the street of Esteve Baudoin (*en la cariera d'Esteve Baudoin*), and the last in the street of Malcohinat (*en la cariera de Malcozinnat*). Elsewhere we find mention

---

[52] ADBR B 812.

of the street of the Dyers (*cariera dels Tenchuriers*) and the street of the Filosas (*cariera de la Fillosas*). The distinction suggests that these houses had not yet been "mapped" onto islands. By the early fourteenth century, as we have seen, council officials were linguistically incorporating isolated houses into islands, and by the mid-fourteenth century the process was essentially complete. The names of islands could change on a regular basis, since they were living entities. Such changes were carefully noted to allow future notaries working for the royal curia to trace obligations, should such an effort ever be necessary. In a record from 1405, the notary, Esteve Chaulan, was especially conscientious in this regard.[53] We learn, for example, of an island located on the Greater street of the Jerusalem "that used to be called [the island] of Nicolau Brasfort and is now [the island] of Johan Bonafos." Another island, once named for oar-makers, had become the island of Dominic de Scalis and Antoni Pomier, the fisherman, and a third, once named for the Blessed Mary of Humility, had become the island of Bertran Lombart, mariner.

Throughout the thirteenth and the first half of the fourteenth century, the entries in these registers rarely named the streets on which a given house was located. In 1301, for example, houses were grouped into islands, and the abutment clauses gave the names of adjoining proprietors or, at best, an unnamed street:

> Next, Johan Vassal holds under the said direct lordship a certain house abutting the houses of Peire Jordan and the shore of the sea. Next, Peire Jordan holds under the said direct lordship a certain house located in the same place and abutting the house of Guilhem Audoart.[54]

No street was given a name in this register.

By 1377, we begin to see the first signs of a change in practice, for although surveys of royal holdings in Marseille continued to organize properties by means of islands, they began for the first time to include named streets in abutment clauses.[55] Two houses belonging to Johan Johan and located in an island named after the Changers (*insula Cambiorum*) were located "in the street of Tomas Luques, abutting the house of Bernat Isnart and Jacme Mercier."[56] Such references were still the

---

[53] ADBR B 1177.
[54] ADBR B 1936. *Item Johannes Vassalli tenet sub dicto directo dominio quandam domum confrontatam cum domibus Petri Jordani et cum ripa maris. Item Petrus Jordani tenet sub dicto directo dominio quandam domum scitam ibidem confrontatam cum domo Guillelmi Audoardi.*
[55] ADBR B 831.
[56] Ibid., fol. 12v.

exception, but reveal a growing awareness that streets are a valid cartographic mechanism for fixing the location of houses. The area around
Negrel street is especially distinctive for the number of streets used.

By 1405, a royal record of rents organized by means of islands was
consistently naming streets in abutment clauses. For example, a property
located in the island of Antoni Bonfilh was described in the following
way:

> On the year and day above, Peire Bertran, a laborer of Marseille, on his
> oath acknowledged to the receiver and to me the said notary, stipulating
> and receiving as above, his ownership of the fourth part of a certain house
> of his located in the said street of the Upper Tannery, abutting on one side
> the house of Peire Gombert, and on another the house of Julian Tacil, and
> behind with a house of Victoret de Massello, and in front the said public
> street.[57]

Occasionally the island was also named in site clauses, but in such cases
the street almost always took priority, as in two houses "located in the
said street of the Upper Tannery in the said island of the said Antoni
Bonfil" (*sciti in dicta carreria Blancarie Superioris dicte insule dicti Antoni
Bonifilii*).[58]

Once streets were recognized it was only a matter of time before the
record itself was fundamentally reorganized. An account book from 1412
organized receipts, as usual, primarily by means of islands, but some of
the names of the islands reveal changing toponyms, and in a few cases
islands weren't used at all, as the examples reveal: (1) the island of Antoni
Bonfil near the old gate of Lauret, (2) the island of Giraut Lort called
the street of the Tannery (*appellata carreria Blancarie*), (3) the island of
Peire de Berre, now Johan de Geminis, street of the Tannery, (4) the
street of St. Martin, Jewish fountain, (5) the Market plaza beyond the
street of St. Martin near the gate (*post carreriam sancti Martini iuxta portale*),
(6) the island of Nicolai Novel, street of Castilhon behind Negrel street.[59]
Particularly interesting is a toponym in which the street has replaced the
island: "the street of the Palace in front of the island of Bernat de
Conquis."

---

[57] ADBR B 1177, fol. 35v. *Anno et die predictis Petrus Bertrandi laborator de Massilie medio iura-
mento recognovit dicto rationali et michi dicto notario stipulanti et recipienti prout super quartam
partem cuiusdam sui hospicii sciti in dicta carreria Blancarie superioris confrontatis ab uno latere cum
domo Petri Gomberti et ab alio latere cum domo Juliani Tacil et retro cum domo Victoreti de Massello
et ante cum dicta carreria publica.*
[58] Ibid., fol. 35r.
[59] ADBR B 1946, fols. 7r–12v.

The office of the *clavaire* was filled on a rotating basis, and the practices of each official could and did vary within certain customary limits. As a result, the change in cartographic language did not proceed at a sedate and regular pace. In a similar listing of receipts from 1439, for example, the clavaire replicated fourteenth-century insular toponymy with few changes.[60] Yet by the middle of the fifteenth century, usage of streets in the accounts of the *clavaire* was becoming more and more frequent. In an account book from 1441, the *clavaire*, listing the entry fees received from ten houses newly sold, used the language of islands on only one occasion; all the others referred to streets. We see not only familiar streets, like Negrel street and the street of the Carpentery, but also appellations that were, for the *clavaire*, rather new. One house was situated in "a street called the street of St. Jacques of the Swords, in Cavalhon"; another "in Rocarbarbola, on a street called of the Gentlemen"; a third "on a street called the Lancery." The exception, significantly, was a house located "in the island of the fountain of the Hospital of St. Antoine," which was located within the largely unmapped area of Cavalhon.[61]

In a different type of register that recorded the state of royal rights in Marseille in 1449, we see a more remarkable transformation.[62] First, Johan Tomas, the notary who kept the register, completely abandoned the practice of organizing acknowledgments of rents owed under insular rubrics, though Johan was certainly aware of islands and made an effort to include them from time to time in site clauses even if such islands were not the fundamental cartographic principle of the register. Second, unlike all previous registers, which began with islands located on the eastern end of the harbor and gradually worked their way westward before turning to islands located away from the port, this register began at the very western end of the crown's holdings along the quay and worked its way eastward. Beginning with a house owned by Jacme Olivier, which Johan Tomas identified as being located on the street of the Oarmakers, the record moves to a house located in the island of the Oarmakers and then to a house located "along the quay or in the island of the Oarmakers."[63] Shortly afterward we pass over a small alley and encounter a house situated "on the quay, that is to say in the Panataria or on the street of the Oarmakers." Three site clauses naming only streets then occur, followed by a site clause listing both street and island, "located in Panataria street and in the island of the Oarmakers." All these houses, in fact, were

---

[60] ADBR B 1948.
[61] ADBR B 1949, fols. 13r–15r.
[62] ADBR B 836.
[63] For these examples see ibid., fols. 1r–3r.

located one after another, touching the quay on the south and the street known as Panataria street or as the street of the Oarmakers to the north. The varying site clauses may reflect the cartography of his informants, some of whom used the language of islands because they knew royal agents usually preferred islands. Whatever the reason, mention of islands was clearly fading.

The reliance on islands, moreover, drops out progressively as we go eastward along the port. Either his interlocutors were using the language less and less or the notary himself was finding it less and less useful to continue to cleave to the record-keeping style of his predecessors as he traveled, figuratively or in person, along the quay. One of the most interesting features of the document is how the notary very carefully described each of the nineteen transverse alleys that divided the long block into its constituent islands. These alleys, although demarcating islands, had rarely been noticed in prior crown records, and here an agent of the crown was becoming aware of these alleys as legitimate and nameable geographic entities for the first time. He did not actually name them; rather, he assigned a number to each, identifying not only where the transverse alley came from and where it went but also what purpose it served (draining rainwater or flushing out the fishmarket), what it looked like (open, partially covered by an archway), and, in one or two cases, both how long it was (e.g. six and a half *canna* or forty-five to sixty feet) and how wide (one *canna*, two to three feet). This is rational-legal bureaucracy *avant la lettre*.

As we move away from the houses along the port, rubrics resurface, as if older habits had begun to reassert themselves. These rubrics sometimes organized houses by streets, sometimes by islands, sometimes both. Islands, however, no longer stood independent of their surroundings; instead, they were carefully identified and located, often according to compass points, as the "island of Franses Noe next to the old gate of Lauret on the southern side of the old wall [with] the Tannery toward the east."[64] One rubric was listed as "the street of the Johans which used to be called the island of Lois Lort." Following this, a different hand, using different ink, has added "on the eastern side" in the margin, a clarification indicating where the houses identified under the rubric were located in relation to the street.[65] Here, then, streets and compass directions were beginning to replace islands, and in the process the older habit

---

[64] Ibid., fol. 30r. *Insulla Francisci Noe iuxta portale anthicam portale Laureti a parte barii veteris versus meridiem* [ink blot] *Blancarie versus orientem.*
[65] Ibid., fol. 35v.

of identifying islands by means of memory was giving way to rational-legal bureaucratic cartography. In this record, where islands are found, they are never identified by means of a genealogy of their former names, except in this last case where the island of Lois Lort, as a rubric, had been replaced by a street.

In a register of accounts from 1465 written not in Latin but in French—likely a product of the close ties between Anjou and Provence that developed during the reign of King René—we find that the language of islands had not totally vanished. Nineteen of the rubrics in this register grouped properties according to island, and twelve according to street or plaza.[66] Hence, French-speaking administrators were not adverse to borrowing the cartographic categories already in use in Marseille. Yet by the sixteenth century, when Provence had entered into the French realm, all island-talk had vanished entirely from crown records.[67] French royal bureaucracy may have played some role in this change, but clearly the process had begun earlier.

A similar transformation took place in the episcopal curia. In the 1423–1425 register, the language of islands still dominated curial practice.[68] In 1470, in a register of acknowledgments, there are no rubrics; site clauses list either streets or, in many cases, simply the ungainly district Cavalhon.[69] Another register in French that lists successive owners of houses was begun in 1471 and was kept up-to-date through the sixteenth and into the seventeenth century. It *did* use rubrics and *all* the rubrics were names of streets.[70] Many of the streets did not have names, since the insular template as formerly used by officials of the episcopal curia hindered the development of street names, and apparently kept on doing so. An episcopal register compiled between 1670 and 1676 returned briefly to the language of islands, although the islands were not named. It does not seem that the practice was in any way a holdover from past curial practice, but instead an experiment with a possibly more rational system of mapping.[71] Curiously, this register also includes the earliest graphic mappings of lordship I have found in episcopal records.[72] Another seven-

---

[66] ADBR B 1952.
[67] See, for example, ADBR B 863 (dated 1538), B 871 (dated 1547 to 1550), B 881 (dated 1560), and B 859 (dated 1562).
[68] ADBR 5G 126.
[69] ADBR 5G 128.
[70] ADBR 5G 135. See also 5G 138 (dated 1510s to 1560s) and 5G 142 (dated 1512 to 1557) for comparable examples.
[71] ADBR 5G 156.
[72] There are nineteen maps beginning on 130. The earliest episcopal map, dated 1638, can be found in ADBR 5G 170, liasse 170.

teenth-century register also used islands on a few occasions. For example, it resurrected two mid-fourteenth-century islands, the island of Fulco Sardine and the island of Durant Barbier.[73]

Both the episcopal and royal curias were responding to a sea-change in the cartographic imagination of lordship that can be dated to the mid-fifteenth century. The arrival of French administration in the 1480s and the French language, some several decades before, seem to have nurtured the change without having inaugurated it. We are left with the task, then, of explaining why it happened. One answer is obvious. These curial documents were usually written by public notaries whose own cartography was influencing the bureaucratic practices of the curia.

A different sort of explanation resides in the possibility or even the probability that an island once recorded a social fact of major significance, such as a noble or patrician enclave like the *alberghi* of Genoa. In the thirteenth and fourteenth centuries, eponymous individuals were sometimes great noblemen or rich aspiring merchants, some of whom, like the knight Johan Athos, owned many properties on the island in question.[74] Early records list such significant names as Augier de la Mar, Johan Vassal, Guitelm de Mari, the Carbonel family, the Vivaut family, and Johan de Serviers. It is easy to imagine that such men exerted a good deal of patronage over their islands. As an oligarchic and nominally pro-Angevin political order emerged over the course of the late thirteenth century, it is also easy to imagine why Angevin officials might have acknowledged the power of these men in the scribal cartography they themselves were just in the process of inventing. It was a cartography that eliminated the powerful and competing cartography of artisanal vicinities and the political order with which this cartography was associated. We can better appreciate, therefore, why artisans and lesser merchants, those who lost the most in the years after the Angevin takeover, should refuse to adopt the language of islands in their own imaginary cartography. Artisanal cartography, as we shall see in chapter four, centered on streets and open spaces, not introspective blocks of houses.

Given the absence of any records prior to 1264 this is entirely speculative. Yet it may help explain a significant change in the language of islands, at least in crown records: by the late fourteenth and early fifteenth centuries, noblemen were often being replaced by lesser figures as the

[73] ADBR 5G 163bis. See the list of rubrics found in the tenth quire. This register cannot be dated precisely.

[74] For this man, who owned six houses on an island that bore his name, see ADBR B 1940, fols. 177v–178r.

name-giver. In a crown record from 1405, for example, islands were being renamed after a fisherman, a barber, a notary, three mariners, a merchant, two fishermen, three laborers, a butcher, a mason, and a currier.[75] Rarely did these individuals own more than a single house in their islands, and it is unlikely that these men exercised any formal patronage over their neighbors. It is possible that the eponymous individuals were serving fiscal or political ends, perhaps acting as intermediaries between the curia and the populations of their islands and thereby protecting the interests of the royal curia. Certainly the role of islanders (*illerii*) in council fiscality was coercive by nature. This may help explain why names of islands were turning over so rapidly by the late fourteenth century. Islands may have been named after their semi-official resident tax supervisors, and the names changed as the "office" was taken up by other men. Whatever the circumstances, it is clear that the names of islands in fifteenth-century crown records were not recording the identity of the island's most noble or honorable individual.[76]

Thus, the island may have been gradually abandoned because it was associated with a fading patronal political order. But the island also had ecclesiastical connotations that may have contributed to its loss of favor. As noted earlier, ecclesiastical cartography routinely carved Christian space into a hierarchy of districts, from the episcopal see to the parish and ultimately to the island. This was a cartography inherited from antiquity and had become thoroughly Christianized in the intervening centuries. We have already seen how the parish was wholly ignored in the cartographies characteristic of fourteenth-century Marseille, and it may well be the case that the island, equally associated with ecclesiastical cartography, was similarly disfavored. One possible explanation for this resides in historical memory, for during Marseille's brief communal phase, in the first half of the thirteenth century, the bishop of the upper city was one of the lower city's major adversaries. Anything associated with the bishop, including ecclesiastical cartography, may well have carried a stigma into the fourteenth century—the union of the former upper or episcopal city and the lower city, after all, did not take place until 1348. Moreover, ecclesiastical cartography is particularly easy to stigmatize, since the claim to define space inevitably carries with it connotations of

---

[75] ADBR B 1177.

[76] The episcopal curia, in contrast, had fixed many names by the middle to late fourteenth century: a number of the islands that show up in the few extant fifteenth-century records bear the names of people long dead. Compare, for example, ADBR 5G 126 (dated 1423–1425) with ADBR 5G 114 (1353–1359). Of the sixteen islands named after individuals in 1423–1425, thirteen were holdovers from the 1350s.

overlordship and control. The evidence is lacking, but the argument, nonetheless, is plausible.

The abandonment of islands entailed an important transformation in the process of identification which was the abandonment of genealogical strategies in favor of mechanical ones. Early on, islands were rarely identified by means of fixed boundaries, compass directions, or other mechanical devices. Instead, as mentioned above, they usually took their name from one of the leading residents of the islands. Of the forty-six islands named after men or women in the register of 1264–1268, thirty-seven bore the names of a current resident.[77] In a more detailed register from 1377, there are thirty-one islands named after individuals. In twenty-five cases, the individual lived on the island in question, and in a further five the rent was paid by a widow, a son, or a daughter. Only one island bore a name that had no obvious relationship to a living inhabitant.[78] This being the case, the names could turn over with great rapidity, in most cases every generation or so, and more rapidly by the later fourteenth and early fifteenth centuries. Hence, crown records often kept careful track of the prior names of the islands in order to facilitate record keeping.[79] This was a record-keeping strategy akin to the making of a genealogy; thus, in the same way that curial officials kept careful genealogies of successive proprietors, so too they kept genealogies of insular names. It was a cartographic strategy that based the imaginary map of the city on significant chains of individuals.

In the crown register of 1447–1449, as we have seen, the curial notary took up the language of streets and accordingly began to reformulate the fundamental map of royal cartography. In so doing, he introduced the mechanical language of compass directions, and began to imagine the city map in much the same way that portolan charts were imagining the world. Compass directions were well known to notaries and figure, for example, in thirteenth-century notarial formularies.[80] As it happens, they were not favored by the notariate of fourteenth-century Marseille, although they were becoming more common by the fifteenth century. The point is that curial notaries and curial officials did not turn to such impersonal and mechanical methods of identifying property sites just because they were more rational. They began to use these methods in

---

[77] Ibid.

[78] ADBR B 831.

[79] ADBR B 831 (1377) and B 1177 (1405) provide the most detailed examples of this practice.

[80] See Salatiele, *Ars notarie*, ed. Gianfranco Orlandelli, 2 vols. (Milan, 1961), 2:229, and Bencivenne, *Ars notarie*, ed. Giovanni Bronzino (Bologna, 1965), 38.

part because the genealogical method was no longer so useful in the changing social world of the fifteenth century where noble or patrician lineages were increasingly less likely to be seen as significant features of the social landscape.

The language of islands was a language that never caught on outside episcopal and royal circles. The public notaries of the fourteenth century used it with obvious reluctance, and just about everyone else—nobles and merchants, artisans and laborers, men and women—avoided it entirely, to judge by identity clauses and other bits of evidence that convey vernacular ideas about cartography. Hence, despite the fact that curial officials had great opportunities to become the semi-official city cartographers—since the claims of lordship were repeated every year, curial officials may have been present at more cartographic conversations than the public notariate—their language clearly seemed too bookish, too fiscal, too divorced from usage, and, most important, too suffused with relations of power to become the norm. Popular resistance to seigneurial cartography helped ensure that a simultaneously developing notarial cartography would become the civic norm. Over time, seigneurial officials, although clearly conservative by nature—since record-keeping is far easier when the format of the record remains constant over years and decades—gradually aligned themselves with the general trend as their own cartography departed increasingly from a civic norm.

CHAPTER FOUR

Vernacular Cartography

Between 1300 and 1666 there were comparatively few changes to the infrastructure of the old city. A few open spaces were created by tearing down buildings. The ancient Jewry lost its identity and became colonized by Christians, who finally mapped its mysterious spaces, and the Hotel de Ville was built on a single block near the port. More significant changes occurred in 1666, for royal interest in using the port as a military installation led to plans for improving the *triste aspect* of the city.[1] Most of this development, however, took place outside the old walls, first on the hitherto sparsely inhabited southern shore of the port, later in the suburbs to the east and north, as Marseille, accommodating its growing population, burst the confines of the medieval city and strove to imitate the airy spaces and rectilineal street patterns already in favor in eighteenth-century Paris.[2]

The architectural fashions of the Second Empire took Marseille as they took Paris, and this time the old city was not spared. In the spirit of Baron Haussmann, the rue Impériale, later renamed the avenue de la République, was blasted in a straight line through the eastern edge of the old city by the engineers Joseph Châtelain and André Barneoud. Commenting on the extensive photographic record of the destruction put together by the photographer Adolphe Terris in the early 1860s, the local

---

[1] The expression is Pierre Lavedan's. See his *Histoire de l'urbanisme: renaissance et temps modernes*, 2d ed. (Paris, 1959), 429–32.

[2] On eighteenth-century debates over city planning, see James Keith Pringle, "The Quiet Conflict: Landlord and Merchant in the Planning of Marseille, 1750–1820" (Ph.D. diss., Johns Hopkins University, 1984).

historian Bruno Roberty, still moved by the event in the mid-twentieth century, observed: "This enterprise brought about (1) the demolition of 1100 houses; (2) the excavation and removal of 1,200,000 cubic meters of earth; (3) the construction of 3,300 meters of drains; (4) the paving of 30,000 meters of both main streets and side streets; and (5) the fabrication of 5,000 meters of sidewalk."[3] The angry sentiments hidden under the matter-of-fact tone of voice were made explicit elsewhere in his writings where he complains about "this mad desire to change, to rip out every existing thing."[4] It was this nostalgia for things past that inspired Roberty to undertake the monumental task of re-creating a street map for Marseille around the year 1423—the year of the great sack of Marseille by the Aragonese—a project to which he devoted much of the last thirty years of his life.[5]

In point of fact, however, the partial destruction of the medieval city in the nineteenth century accomplished in a material sense something that had already been achieved by the eighteenth century in a linguistic sense, for if the infrastructure of the intramural city changed little from 1300 to 1800, the same cannot be said for the city's linguistic cartographies. The period saw a wholesale turnover of names and the elimination of the cartographic templates not just of island but also of vicinity and landmark from official discourse. As I will argue in this chapter, vicinity and landmark were the chief elements of vernacular linguistic cartography, and although the templates appear in site descriptions in fourteenth-century notarial casebooks, they were used much more often in address clauses by artisans, merchants, retailers, service trades, professionals, laborers, and fishermen. The vicinal template assigned names *not* qualified by "street" to small, neighborly areas that encompassed a segment of a longer street, adjoining alleys, and possibly the houses as well. Some examples are Corregaria, Hill, Auction, Spur, and Jewelery. In fourteenth-century notarial documents kept in Latin, they usually appear in the nominative or accusative. In Provençal, a language with no declensions, they are prefaced by prepositions such as *en, en la,* or *a,* according to the shape or function of the space designated.

Vicinity was a prominent marker of identity in the vernacular linguistic community made up of speakers of Provençal in fourteenth-century Marseille. Two forms stand out. One was based ultimately on a landmark or other feature of the landscape, such as the name of a house or tavern,

---

[3] ADBR 22F 100, deuxième partie, XI, "Les photographies du IIe Empire."
[4] Ibid., X, "Les avatars d'une place publique."
[5] He first began his genealogical work in the early 1920s, and his interests evolved and grew considerably in the years before his death in 1950.

that had developed into a convention. The second was based on readily identifiable centers of artisanal production or of retailing. As historians and geographers have frequently observed, identification with quarters responded to the perceived needs either of trade groups or of kin groups. Given this situation, the disinclination of Marseille's notaries to acknowledge this identity template is especially noteworthy.

The Templates of Vicinity and Landmark

Notarial cartography, by and large, was a Latin cartography, for all notarial casebooks from the period were kept in Latin and usually used Latin words and expressions to describe the physical geography of the city. From time to time, however, notaries slipped into Provençal in personal identity clauses and even in site clauses. When they did so, their syntax changed in subtle but significant ways. In May of 1358, for example, a woman named Sileta Enrica ratified a house sale made by her husband, David, and the notary who drew up the act, Peire Giraut, identified the house as *sita in civitate Massilie in carreria dicta la Veyraria Viella*, which translates literally as "situated in the city of Marseille in a street named the Old Glaziery." In this case, the Old Glaziery is not embedded within a prepositional phrase. This was unusual, since streets in notarial site clauses were almost always given in the genitive. The difference is that "Veyraria Viella" is a Provençal expression, and the word *dicta* ("named") signifies the transition from Latin to Provençal. The Provençal expression here does not include a cognate of *carreria* or even the other Provençal word for street, *rua*. In several other notarial site clauses from the mid-fourteenth century a similar situation holds, as in areas called, within Provençal, the Enclosures, the Figgeries, Malausena, Spicery, Arches, and several others.

Notarial consciousness of the distinction between Latin and Provençal in matters of naming city spaces seems more acute in phrases found in fifteenth-century casebooks. Here, we find a plot located "in a street called, from of old, the Corrageria," and a house located "in a street called, from of old, the Goldsmithery."[6] Another notary described a house "in a street called, in the vulgar tongue, lo Mal Cozinat del Lauret,"[7] and a third notary located a house "in a place called the Spur" (*in loco dicto l'Esperon*).[8] In other casebooks we find "the Wide Furriery," "the Gavot-

---

[6] ADBR 351E 408, fols. 112r and 156v.
[7] ADBR 351E 367, fol. 51r.
[8] ADBR 351E 344, fol. 161r.

tas," and "the Fishmongery."[9] More frequent use of the words "dicta" and "vulgariter" in the mid-fifteenth century suggests a greater consciousness of the gap between Latin and Provençal cartographies. At times, these Provençal expressions were also introduced with the Provençal equivalent of the genitive, namely, the preposition *de*. Thus, in 1453 we hear of a house and little garden "in the alley near a street called, in the vulgar, the Almoner's," and a notary who at one point in his casebook described a site as "a place called the Spur" elsewhere described the location as "a street called 'of the Spur'."[10]

Judging by these indices, the Provençal spoken in fourteenth- and fifteenth-century Marseille characteristically used a cartographic grammar distinct from that of Latin, a grammar in which residential areas were often known as free-standing nouns and not as genitival phrases modifying street, alley, plaza, or any other element of the skeletal framework of public ways. From the evidence just surveyed one could even argue that notaries, who were also speakers of Provençal, used this vernacular cartography in everyday life, and only thought in terms of streets in the legal context of the acts they drew up. Yet the indices are drawn from inherently slippery sources, since they were written by notaries who were writing primarily in Latin. To explore the distinctive cartographic grammars characteristic of Latin and Provençal, we would need to measure the notarial Latin usages that constitute the bulk of the surviving evidence against a source that (1) was kept in the Provençal language; (2) was not written by a notary and preferably not written in the service of the city council or the Angevin crown; and (3) included some sort of geographic nomenclature.

Incredibly enough, there is a document that meets all of these conditions. In 1347 the wealthy merchant, Bernat Garnier, founded a hospital known as St. Jacques de Gallicia. A confraternity associated with the hospital arose simultaneously and scrupulously kept accounts that listed the names of members both old and new, the size of their annual contributions, other sources of income, and expenditures. One of these registers, written in Provençal and covering the period 1349–1353, has survived; its peculiarly angular handwriting is typical of literate merchants or artisans, not notaries.[11] In one place, the scribes were identified as Johan

---

[9] ADBR 381E 107, fol. 71r; 351E 378, fol. 106r; 351E 330, fol. 84v.
[10] ADBR 351E 333, fol. 134r; 351E 344, fols. 136v, 138v, 151r.
[11] The document is ADBR 2HD E7. The document has been paginated by a modern hand; I have followed this pagination. The handwriting is similar to the handwriting found in other mid-fourteenth-century documents kept in Provençal by merchants, such as AM EE 55A and EE 55B, and ADBR 1HD H3, 1HD B102, and 4HD B1 (portions).

Esteve, Enric Astier, and Antoni Dalmas.[12] The prosopographical index tells us that Enric was a baker; Antoni, in turn, was probably a merchant who lived in Corregaria. There were many Johan Esteves in the city; this one was identified in the confraternal register as a resident of the Upper Grain Market and in all likelihood was the laborer identified in episcopal records as a resident of the same place.[13] The officers of the confraternity tended to rotate. Between 1349 and 1353 these officers included, in addition to the men above, the furrier Antoni Simon, the crossbow-maker Jorgi Buenenfant, the shoemaker Itier de Sion, the clothier Guilhem Dalbis, the currier Jacme Donadieu, the cobbler Jacme Johan, the shoemaker Bertran Rostang, and the buckler Peire Arman.[14] These agents of record, in short, were entirely representative of ordinary Provençal speakers.

The 560 different members whose names are inscribed in the register provide a total of 376 addresses. Figure 7 maps out the rough location of all members of the confraternity who provided addresses that can be known according to the street-based Roberty map (a number of addresses, of course, cannot be sited on the map, especially those relying on unknown landmarks). The map reveals that recruitment was fairly broad throughout the city, despite the location of the confraternity in the extreme eastern end of the intramural city. Many of the members of the confraternity also listed their professions or, if women, those of their husbands, and from these we can tell that the confraternity recruited heavily among artisans and service trades, especially among shoemakers, butchers, bakers, and clothiers (Table 4.1). Curiously enough, some individuals who can be identified in the prosopographical index as nobles chose not to use this label when identifying themselves to the confraternal scribe; alternatively, the scribe was reluctant to use the title. The same is true for agricultural laborers, who accounted for at least twenty-nine of the individuals who did not give their professions when identifying themselves. These identifying labels, clearly, do not measure recruitment among artisanal groups so much as they measure the depth of professional pride characteristic of particular groups or the bias of the scribe. One result of these identity preferences is that the confraternity in fact represented a greater spectrum of Massiliote society than a rapid glance at the register would suggest.

The vernacular addresses found in this register are not, strictly speaking, comparable to the site clauses found in notarial casebooks, since they

[12] ADBR 2HD E7, p. 64.
[13] See ADBR 5G 114, fols. 50v–51r.
[14] ADBR 2HD E7, see the listings on pp. 2, 45, 64, and 68.

Table 4.1. Trades represented in the confraternity of St. Jacques de Gallicia, 1349–53

Shoemaker	26	Mason	3	Canvas maker	1
Butcher	24	*Penhedor*	3	Carter	1
Baker	14	Saddler	3	Caulker	1
Clothier	12	Apothecary	2	Cloth-shearer	1
Carpenter	8	Cooper	2	Crossbow maker	1
Fisherman	7	Currier	2	Laborer	1
Candler	5	Cutler	2	Merchant	1
Cobbler	5	Farrier	2	*Pecadort*	1
Goldsmith	5	Furbisher	2	Plasterer	1
Hatter	5	Gardener	2	Priest	1
Inn-keeper	5	Grave-digger	2	Slipperer	1
Smith	5	Locksmith	2	Teacher	1
Pastry-chef	4	*Nacarat*	2	Weaver	1
Barber	3	Tanner	2	Total	177
Buckler	3	*Triquiera*	2		

Source: ADBR 2HD E7.

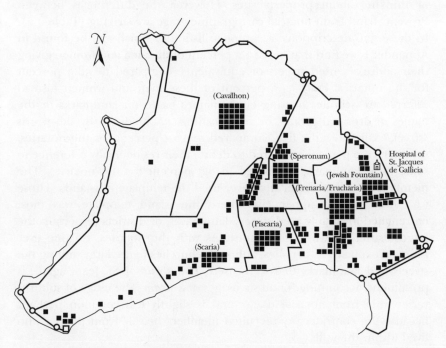

7. Membership in the confraternity of St. Jacques de Gallicia by place of domicile, 1349–1353. Each square represents the approximate domicile of a member of the confraternity.

Table 4.2. Comparison of cartographic categories used in vernacular addresses and notarial site clauses, 1337–62

	N	Streets (%)	Districts (%)	Vicinities (%)	Landmarks (%)
Vernacular addresses	376	13.3	16.2	54.3	16.2
Notarial site clauses	932	58.3	16.6	16.4	8.7

Sources: ADBR 2HD E7; 300E 6; 351E 2–5, 24, 641–45, 647; 355E 1–12, 34–36, 285, 290–93; 381E 38–44, 59–61, 64bis, 72–87, 393–94; 391E 11–18; AM 1 II 42, 44, 57–61.

were meant to identify the addresses of individuals, not house sites. Nonetheless they are the best source we have for exploring the ways in which Provençal speakers described their residential spaces independently of notarial influence. There is no particular reason to think that Provençal speakers would have used dissimilar terms when defining or thinking about property sites. However, the differences between Provençal and Latin notarial cartographic usage are striking (Table 4.2). In these 376 descriptions, a complete listing of which can be found in Appendix 1, we find that only 13.3 percent of the men and women giving their addresses used streets or equivalents, compared to 58.3 percent for the notaries. Fully 54.3 percent of these men and women instead referred to vicinities bearing either names based on landmarks or the names of artisanal groups or retail centers *not* modified by the words "street," "alley," or the like, compared to 16.4 percent for the notaries. Another 16.2 percent preferred to define their residence by reference to some local landmark, compared to 8.7 percent for the notaries. The members of the confraternity never used the template of islands. Those who lived in the upper city where the island template was most entrenched invariably used streets, landmarks, or districts (especially the large district known as Cavalhon). As with the notaries, no one ever referred explicitly to parishes, although churches figure large among the streets and landmarks that were used. The *sixain* of St. Jean, again, is prominent, accounting for all six usages of a *sixain*. The usage of suburbs varies little from notarial usage, and is slightly less common perhaps because the confraternity recruited members heavily from people who lived within the walls.

Streets

Streets were used in some fifty confraternal addresses, indicating that Provençal speakers, though characteristically attuned to vicinities, had the cognitive capacity to take heed of streets. The multiple spellings given to the Latin word *carreria*, which was transliterated into the Provençal of this register as *cariera*, *carriera*, *carier*, *quariera*, and the unmodified *carreria*, reflects not only the phonetic spelling pattern that is characteristic of all Provençal from this period in Marseille but also, perhaps, a lack of familiarity with the word itself.[15] Curiously, streets were often used by isolated individuals. Twenty-seven streets are named in this register and only nine were inhabited by two or more members of the confraternity. Negrel street, a major shoemaker enclave, is a major exception, for it was identified by thirteen individuals as their place of domicile.

If members of the confraternity typically preferred to use the nomenclature of vicinities, why did they use streets at all? Many of the areas described here in terms of the street template, in all extant documentation from the mid-fourteenth century, were never described as anything other than streets, indicating a usage hardened by decades or centuries of practice. Negrel street, for example, was always a street in all mid-fourteenth-century documents, Latin or Provençal, as were several other streets that figure in the confraternal register, such as the street of the Almoner, the street of Castilhon, the street of the Engarian, and the street of Francigena. Many of the other streets listed in the register were also known as suburbs, and several, such as the street of St. Martin, identified areas also known in this register and elsewhere by means of the landmark template (e.g. *denant Sant Martin*). In these cases, there were no vicinity-based names available for adoption by speakers. Put differently, the template of vicinity, at least in the mid-fourteenth century, apparently could not be used to map out all the spaces in the city. Only a few of the streets found in the register were known elsewhere by means of the template of vicinity: these include the street of the Smiths, also known (albeit very infrequently) as the Smithery; the street of the Marquesas, also known elsewhere in the register as the court of the Marquesas; the street of Prat Auquier, more commonly named in the register as Prat d'Auquier (8 usages); and the street of Perier, also known in some records simply as Perier. That these people used the streets when vicinities were readily available in language indicates, I would argue, that the notarial street-based template was beginning to infiltrate the ordinary cartographic grammar of Provençal speakers.

---

[15] The word does not appear in the dictionary of medieval Provençal provided by Pierre Pansier in his *Histoire de la langue provençale à Avignon du 12ᵉ au 19ᵉ siècle* (Geneva, 1974 [1924]).

This process can also be glimpsed in the grammatical construction of Negrel street. The name of the street was sometimes given as "cariera Negrelli" and sometimes "cariera Negrel." The former, in the genitive case, is Latin, and the latter duplicates how Provençal speakers might have normally named the area. Yet Latinate usage (found eight times) was more common in this confraternal register than Provençal (five times), suggesting that spoken Provençal was actually coming to use the Latin expression. There is only one other example in the register of a Latin genitive ending used with street; this was "street of the Smiths" (*cariera Fabrorum*). Normally, the word *cariera* appears with undeclined nouns, sometimes with the preposition *de*, sometimes alone (e.g., *cariera del Perier, cariera de Jaret, cariera Jaume Cancel, cariera Jaret*). Why a notarial and Latinate spelling of the *cariera Negrelli* was seeping into everyday Provençal usage is not easily explained. The street was densely populated by a highly self-conscious group of shoemakers. It was also a relatively long street with a great many houses located along its length. Since long streets inevitably witnessed more property conveyances, the cartography of long streets was pondered more often by notaries as they drew up site clauses. One result is that the long streets in Marseille—such as the New street, the street of the Jerusalem, the street of Lancery, and the street of the Upper Grain Market—were among the earliest to acquire permanent, unchanging names. The shoemakers living along Negrel street apparently adopted the name as their toponym.

The usage was surprisingly powerful, as we have seen, since it penetrated the cartographic custom of the royal curia, more accustomed to using insular talk.[16] The area around Negrel street was among the first to be remapped according to street-based usage in crown records, a remapping that resulted in the oxymoronic rubric, "the islands of the streets of Negrel, of the Galli, of the Almoner, and of Castilhon" (*Insulae carreriarum de Negrello, dan Galli, de Elemosina, et de Castilhone*).[17]

Vicinities

The dominant template in the artisanal cartographic imagination was the template of vicinity. Twenty-four of the vicinities used as addresses by members of the confraternity of St. Jacques de Gallicia were named after artisanal groups or service trades; twenty more took Provençal names not

---

[16] Above, chapter 3.
[17] This rubric can be found in ADBR B 831, dated 1377.

immediately related to a trade group. Some of the Provençal names had no particular meaning. Others, such as Cavalhon, may have been so named because the area was once occupied by immigrants from the town of Cavallion further to the north on the Rhone, but names such as these often outlive their origins and acquire their local meanings over the course of time. Still others were named after some kind of landmark, such as Steps or Pilings, or features of the physical topography of the city, such as Hill. For purposes of classification, they could just as easily have been included under the rubric of "landmark," and indirectly show the close linguistic or cognitive association between vicinities and landmarks. In some cases, the distinction I have drawn between an artisanal vicinity and a landmark vicinity is also a slightly artificial one, a distinction based on time and maturation, not derivation. Such is the case with Cayssaria, which took its name from the moneychanging trade ("Cashery"), Lansaria, probably named for armorers who made lances ("Lancery"), and Corregaria, based on an old word for cobbler. These areas had been dissociated from the eponymous craft groups for so long that the words seem to have taken on new, autonomous significations.

Vicinities were not streets. They were something more. In the most typical pattern, a vicinity was associated with a segment of a longer street, but included within its ill-defined boundaries some of the alleys or streets that fed into the street segment and all the houses located on these streets and alleys.[18] In principle, therefore, it should be possible to find a house site or address identified with a vicinity by Provençal speakers but located, by Latinate notaries, on a lexically distinct street. This is precisely what we do find, and I shall be discussing these revealing examples of lexical disagreement in the next section. The geographical extent of vicinities varies considerably. Some vicinities consisted of nothing more than a single street and its houses. Such was the case with the Carpentery which was located in a narrow, intramural space and was therefore constrained by its physical location to the street that ran down the middle. Other vicinities extended over a much larger terrain, and references to Cavalhon and the "quarter of St. Jean" were so common that these, too, should be seen as constituting a kind of super-vicinity. Most vicinities, however, fell in between these extremes.

---

[18] I am not the only one to have noticed this. One of Marseille's leading topographic historians, Eugène Duprat, noted that "The *Bocaria* is the name of a quarter and of the three or four streets that joined together." See Fernand Benoît et al., *Monographies communales, Marseille-Aix-Arles*, vol. 14, part 3 of *Les Bouches-du-Rhône. Encyclopédie Départementale*, ed. Paul Masson. (Paris, 1935), 107.

More than merely elements of linguistic cartography, vicinities were units of sociability and social identity. This is clearest in the case of artisanal vicinities. To the extent that they were inhabited or used by practitioners of the given craft, these vicinities were units of industrial production and surveillance. But all vicinities appear to have been high-status, desirable addresses. This is made clear by the obvious preference on the part of many confraternal members for seeing themselves as living within clearly defined vicinities. Fruitery, Fishmongery, Jewelery, Goldsmithery, Cobblery, Carpentery, Shoemakery, Furriery, Cavalhon, Spur, Pilings, Corregaria, Hill, Prat d'Auquier, Tripery, Jewish fountain—each of these sixteen vicinities was named as a place of residence by five or more people, and account for 165 or 44 percent of the 376 addresses given in the confraternal register. Only three of the suburbs and one other area, namely Negrel street, were mentioned five or more times by members of the confraternity. Heavy recruitment from these sixteen vicinities surely played a role in their greater-than-average representation, but even this is telling, for, according to this argument, it is primarily in vicinities such as these that social solidarity and vicinal identity was sufficiently strong to make such confraternal recruiting efforts successful.

This stands in sharp contrast to the streets found in the confraternal register. Excluding Negrel street as a special case because of its *de facto* status as a shoemaker vicinity, the twenty-six streets named in the confraternal register averaged between one and two inhabitants per street.[19] In contrast, the forty-three vicinities averaged closer to five inhabitants. Even the forty-three landmarks, which averaged about two inhabitants, were better represented than the streets. This discrepancy could mean that these particular streets had no particular vicinal solidarity, and hence the recruiting efforts of the confraternity met with little success. But it is more likely that many people who gave their address in terms of one of the favored vicinities did so simply because they wanted to be associated with the vicinity, and lived close enough to one of the vicinities to be able to claim it as a place of residence. The Fishmongery (*Piscaria*) is surrounded by "white space" claimed as a place of residence by no one (Figure 7), and the Spur (*Speronum*) is ringed by white space in an arc running clockwise from the northeast to the south. These white spaces by no means show that no one lived in the areas in question.

[19] The *carreria Negrelli* was one of the few areas of the city where a great density of shoemakers did not result in the appellation *Sabateria*. Nonetheless the consistent and seemingly proud usage of the address by shoemakers attests to the sense of shoemaker identity with the *carreria Negrelli*.

Rather, they suggest how the gravitational pull of areas like the Fishmongery and the Spur affected the mental cartography of members of the confraternity.

Vicinal identity was strong because it mattered. Among other things, the memories of neighbors acted as archives containing legally important facts like time of birth and death, title in property, marital status, and so on, information that would later fall into the purview of state archives.[20] Perhaps even more important, neighborhoods were moral and honorable spaces that frowned on gambling and uncontrolled sexuality. We can touch this world only through chance remarks found in pleas or witness depositions in court cases. These remarks reveal a world divided, with affecting simplicity, into good spaces and bad spaces. The good spaces were called *bonae carreriae*, "good streets," or simply *honesta loca*, "good places," in contrast to *inhonesta loca*.

Court cases illustrate the moral sentiments that pervaded these neighborhoods. At some point in the 1330s, a group of shoemakers living on Negrel street began complaining to judges and other officials about the antisocial behavior of a fellow shoemaker, Antoni d'Ays.[21] Antoni, it seems, had lost his wife, Dulcelina, who had left him one day and was no longer to be found in the city of Marseille. Unable to remarry and therefore not having "anyone who could do his service," as he would later put it during the course of the ensuing trial, he acquired a maid named Dulcia.[22] But his neighbors and shoemaker colleagues did not like Dulcia and did not like the way Antoni was carrying on with Dulcia; they told the judge that Antoni was prostituting her. This contravened a statute entitled "On Female Libidinousness," and also a more recent proclamation to the effect that "No inhonest woman or prostitute should dare to live in a good street in the city or its suburbs."[23] As one witness named Peire Razos explained, he considered Dulcia to be:

> a woman vile of body, in that Dulcia lived for a long time with Anthony as his concubine, garrulous and also quarrelsome, overflowing with violence, wickedness, and garrulousness for the reasons expressed above. He added that Dulcia was not worthy to live in Negrel street nor in any other good or wholesome street (*alia bona seu sufficienter carreria*) on

---

[20] See my "Los archivos de conocimiento y la cultura legal de la publicidad en la Marsella medieval," *Hispania: revista española de historia* 57 (1997): 1049–77.

[21] ADBR 3B 41, fols. 164r–185v, case opened 16 Aug. 1340.

[22] Ibid., fol. 164r.

[23] AM FF 165, fol. 2r.

account of the fights that she continually has and had with the ladies and other people of the street, and even on account of the danger to the men living on the street because her carnality sticks out (*propter periculum hominum habitancium in dicta carreria quod posset verisimiliter garnalitate ipsi Dulcie eminere*).[24]

Antoni, of course, did not agree, and sought to prove with his own witnesses that "it is lawful for anyone in Marseille or elsewhere to have a woman as a maid for doing his business, as long as he does not prostitute her in honest streets."[25] As one of the witnesses—Paulet Faber, son of a jurist—put it, "he heard it said that any man not having a wife might keep a woman or prostitute on a good street as long as he does not prostitute her, and as long as the woman does not carnally know other men in the street."[26]

A second example underscores the point. In the late 1330s, four male residents of the vicinity of Malcohinat banded together to try and evict a suspected prostitute, Mathendis Ruffa, under the terms of a statute requiring that "prostitutes living among upright and honest men must leave at their request or be expelled."[27] The men in question—Antoni Laugier, the notary Peire Main, Peire Arnaut, and Martin de Vaquiers—were successful in their lawsuit, and according to the condemnation delivered by the viguier, Mathendis was required "to vacate the house located at Malcohinat near the gate of Lauret and leave the vicinity within three days, or else he would have her evicted from the said house."[28] Mathendis, who described herself as a married woman and hence respectable, did appeal the sentence in October of 1339 and, in fact, there may have been some question either about her guilt or about the real motives of the four men. Still, the existence of the statute and the way in which it was successfully used in this case reveal the powerful associations that could exist between vicinity and morality.

These two cases were among several in which witnesses used the language of "good streets." People were concerned about the reputations of their streets because their own reputations depended, in part, on the moral spaces in which they lived. Thus, to provide their own good reputation—*bona fama*, as it was called—plaintiffs would get their

---

[24] ADBR 3B 41, fol. 167r.
[25] Ibid., fol. 174r.
[26] Ibid., fol. 177; *non habens uxore potest tenere unam mulierem sive meretricem in bona carreria*.
[27] *Statuts*, book 5, 100.
[28] ADBR 3B 805, fols. 12r–15r, case opened 18 Oct. 1339.

witnesses to say how they had never seen the plaintiff in vile, inhonest places. As Lois de Bonils said about a fellow notary, Guilhem de Belavila, on trial for homicide, "he doesn't go to taverns or other inhonest places (*inhonesta loca*) and he frequents churches."[29] The "inhonest places" referred to by Lois were parts of the city frequented by gamblers, drunkards, and ruffians.[30] In registers of the deliberations of the city council we find gamblers sometimes being chased out of certain streets.[31] The Change, we learn in one register, "is from antiquity an honest place," and therefore gamblers are forbidden to frequent it, although the fine print records that nobles and other "good people" are still allowed to play certain kinds of games (*ad scacos et tabulas*).[32] The working and living space of prostitutes was similarly controlled; the urge to regulate prostitutes was probably becoming more intense across the fourteenth century.[33]

Morality mattered primarily because it affected female status and reputation. Catherine Ossa, fearful that her daughter was being seduced into prostitution by a neighbor, Jauma Jugueta, gave Jauma a bitter tongue-lashing—calling her, among other things, "a horrible Hebrew truant... who will make a prostitute of my daughter"—and was fined the massive amount of twenty royal pounds for this expression of maternal concern.[34] Respectable people like Catherine Ossa did not want their living spaces tainted by the scandal of prostitution, in part because people were closely identified with the areas in which they lived. They voluntarily made this identification themselves. In the register of the confraternity of St. Jacques de Gallicia, when people bothered to identify themselves at all, they typically chose either trade or domicile, rarely both. Like trade, place of

[29] ADBR 3B 825, fol. 166r.
[30] See, for example, ADBR 3B 62, fol. 115r–v; the plea attempts to discount the hostile testimony of a man named Andrieu Scoffier who hung around with other men of low standing in "inhonest places . . . and other most vile places" (*locis inhonestis . . . et locis vilicimisis*). Another representative case can be found in ADBR 3B 826, fol. 52v; we meet someone named Guilhem Jordan accused of being "a vile man, frequenting taverns and prone to drunkenness."
[31] AM BB 20, fol. 37v.
[32] AM BB 26, fol. 21r.
[33] See AM FF 169, fol. 18v (1365–66), a cartulary of public pronouncements which makes reference to an injunction of the queen ordering prostitutes "not to enter any house in any street of this city that is part of an honest vicinity (*alicuius carrerie huius civitatis honeste vicine*)." Leah Otis shows that towns in Languedoc were increasingly likely to desire to control prostitution by the middle and late fourteenth century; see Leah Lydia Otis, *Prostitution in Medieval Society: The History of an Urban Institution in Languedoc* (Chicago, 1985), 77–88.
[34] AM FF 519, fols. 61r–v, cased opened 9 Aug. 1341; *orra truanda esbraigua que . . . farai putan de ma filha.*

domicile was a potent indicator of identity, and the vicinity, with all its social connotations, was the preferred format.

Landmarks

The men and women of the confraternity were not only much more inclined than the notaries to use the template of vicinity; they were also more inclined to refer to addresses by means of a landmark, used in 16.2 percent of the addresses given in the register. For the sake of consistency, I have classified landmarks according to the same principles I used in chapter two: with some exceptions, the category of landmarks consists of religious edifices, cemeteries, significant civic buildings, ovens, markets, fountains, gates, and notable individuals. Thus, we find people living "before St. Louis," "at the Preachers," "at the Palace," "at the Gentleman's Oven," "at the Old Market," "at the Gallican gate," and "at Lauret." The list includes eleven addresses defined merely by reference to a local notable, such as "before the house of Sir Peire Austria," "next to Folco Audebert," "before Guilhem Estaca, mason," and "before Marques Malet." The relatively large number of addresses based on notable figures is interesting, because it shows that the template of neither street nor vicinity thoroughly dominated popular cartographic discourse. Usage was casual, since some people just made up an address as they went along and did not bother about conforming to any norm. Most of the notables linked in this way to addresses were members of the confraternity, suggesting that the man or woman who was giving his or her name and identity to the scribe had said something like, "Well let's see, I live right across the street from Guilhem Estaca's house."

Vernacular Classification

For the purpose of analysis I have imposed certain principles of classification on the vernacular addresses found in the register of the confraternity of St. Jacques de Gallicia. The category of "street" is relatively unambiguous, since it is marked by the word "cariera."[35] Given this standard, the distinction between landmarks and vicinities is an artificial one, since the categories themselves are not denoted by a term or expression equivalent to the word "cariera." To complicate the matter, many entities

---

[35] The only exception to this is Quay, which appears in the register simply as *Riba*. The public notaries, however, clearly used "Quay" (*rippa portus*) as a kind of public space, much like street, plaza, or alley, and since I included the quay in the category of public spaces in chapter 2, consistency demands its inclusion in the same category here.

I have called vicinities took their names from landmarks. The name for the Spur, for example, was probably derived from a house or tavern sign; alternatively, it may indicate a spur jutting out of the hill. The Spur, thus, could be classified as a landmark, although in practice the name had long since outgrown its derivation and had become conventional. Similarly, the Jewish fountain was identified as an address by no less than twelve members of the confraternity. More than just a convenient nearby fountain, it had become a linguistic convention, and hence a neighborhood, a place where people lived or, in other words, a vicinity. Yet how do we measure conventionality? Is there any principle of classification that was native to Provençal usage?

Indeed there was such a principle of classification, based on prepositions.[36] Between "next to Folco Audebert" and "in the Fishmongery" there is a considerable difference, since one is clearly a landmark, the other clearly a place "in" which one lived. Members of the confraternity of St. Jacques de Gallicia usually described themselves as living "in" artisanal vicinities and, for that matter, "in" streets as well (Provençal: *en, en la,* and *en lo*). Vicinities named after trades, like the Fruitery, the Jewelery, and the Upper Drapery, were easily imagined as spaces with boundaries or as encompassing social entities, and these also always took the preposition "in." A few vicinities not named after trades, like Cavalhon and the court of the Marquesas, again took "in," and therefore were imagined as enclosing spaces. In contrast to this, vicinities originally named after landmarks typically took the preposition "at" (Provençal: *a, a la, als*) and did so when the entity from which they took their name could not be easily imagined as an enclosing space; typical of this pattern are the Hill, the Jewish fountain, and the Pilings. It is easy to see why a hill, a fountain, or pilings might not readily lend themselves to the preposition "in." Members of the confraternity also referred to the *sixain* of St. Jean on six occasions and used "at" in so doing (*a Sant Johan*)—as would a speaker of modern French. The distinction is captured in the expression used by mistress Antoneta Mouniera, who explained to the confraternal scribe that her domicile was located "at St. Jean in the street of the Figgery" (*a Sant Johan en la quariera de Figier*).[37]

As linguists note, these are not trivial distinctions.[38] Claude Vandeloise

---

[36] Cognitive linguists argue that basic cognitive patterns are revealed in the use of parts of speech and metaphors. Particularly helpful is George Lakoff and Mark Johnson, *Metaphors We Live By* (Chicago, 1980); Stephen Pinker, *How the Mind Works* (New York, 1997), 352–58.

[37] ADBR 2HD E7, p. 42.

[38] See, for example, Sally Rice, "Prepositional Prototypes," in Martin Pütz and René Dirven, eds., *The Construal of Space in Language and Thought* (Berlin, 1996), 136.

remarks that the preposition *à* in modern French "functions essentially to locate the target with respect to the landmark."[39] The functional relationship between the target and the landmark captured by the preposition *dans* is rather different. In the case of an expression like "the fish is in the hand," Vandeloise observes, a large fish can in no way be considered to be contained within a containing hand. In cases like this, the "landmark" (the hand) is exerting a certain amount of force on the target (the fish), thereby enabling use of the preposition *dans*. We can easily extrapolate the argument to the social force exerted by a street or vicinity. Given the regularity of prepositional usage in modern languages, it comes as no surprise that there is remarkable consistency in the pattern of prepositional usage in both Latin and Provençal sources from medieval Marseille: landmark vicinities typically take the preposition "at" and vicinities named after artisanal or retail trades typically take "in." The Fishmongery is the only vicinity I have found that could take both: residents of the area used "in" and "at" in roughly equal proportions.

Whatever the differences between "in" and "at," both marked conventional features of the landscape. The conventional nature of these prepositions stands in sharp contrast to the ad hoc quality of other types of spatial prepositions, such as "next to," "in front of," "below," "above," or "near" (*costa, denant* or *davant, desot, sobut,* and *prope*). These typically denote non-conventional landmarks. In the case of notable individuals, clearly one does not live "in" or "at" them, so we find the mason Johan de Branges living "before Sir Peire Austria" (*denant sen Peyre Austri*), and in separate entries the couple Borga Blanqua and Peire Blanc both described themselves as living "next to (*costa*) Peire Arman." As it happens, churches generally took the preposition "at," suggesting that they could be classified as landmark vicinities, but they could also take relational prepositions as well, indicating that some people saw them as landmarks. Johaneta Aymara, we learn, "is at the Franciscans, in the front" (*esta als Frayres Menos denant*), Johan Nicolau lives "before St. Martin" (*denant St. Martin*), and Johan Borguonhon lives "under St. Esprit" (*sobut St. Esprit*).[40]

There is, then, a very simple rule that we can use to distinguish between

[39] Claude Vandeloise, *Spatial Prepositions: A Case Study from French*, trans. Anna R. K. Bosch (Chicago, 1991), 160. See also Paul Bloom et al., eds., *Language and Space* (Cambridge, Mass., 1996).

[40] ADBR 2HD E7, pp. 42, 71, 76, and 113.

one kind of a living space and another. Use of "in" or "at" indicates an area that has become a conventional category of vernacular linguistic cartography, whereas the use of relational prepositions indicates a landmark in casual usage. For example, Guilhem Estaca explained to the confraternal scribe that he lived "before the oven of the street of the Almoner" (*denant lo forn de la quariera dell Almorna*). The address was clearly an impromptu one, for people customarily used the street of the Almoner alone as their address. In contrast, Peire de Menreza lived "at the Gentleman's Oven" (*al Forn Danprodome*).[41] This was a well-known oven that was known either as a street that one lived "in" or as a landmark that one lived "at." In becoming an oven "at" or "in" which one lived, and not simply an oven "before" which someone lived, it had come to symbolize a larger social space. Certain ovens, evidently, underwent a kind of ontological and linguistic evolution and became synecdoches for larger vicinities, whereas others remained just ovens across the way. The same would have been true for fountains or any other landmark.

In theory, a history of changing prepositional usages would allow us to track the historical process whereby landmarks like this were gradually incorporated into linguistic conventions. In practice, this would be a difficult history to write, at least with the sources extant from medieval Marseille. The vast majority were written by notaries, and therefore do not necessarily reflect the common usage of Provençal speakers. Latin notarial street-based usage, moreover, reduces all lexical terms to their genitival form, creating a universal prepositional format ("the street *of* the Gentleman's Oven") that can no longer reflect the prepositional subtleties of vernacular linguistic cartography.

Notarial Translation

In their frequent use of indefinite landmarks and of vicinities of all types, the members of the confraternity of St. Jacques de Gallicia were not tempted, as were the notaries, to translate all property sites or addresses into streets. The scope of this notarial act of translation, described in chapter two, comes out more clearly here, for the addresses found in this Provençal register are the closest approximation we have to what people actually *said* in the fourteenth century when asked by a notary to name a place of domicile or identify a house site, and we are in a better position to measure what the notaries did with this

---

[41] Ibid., pp. 36 and 69.

information. To judge by the confraternal register, ordinary men and women, confronting a request that they identify a property site, would have normally—around 70 percent of the time—answered by identifying the site in terms of landmarks or vicinities. Yet only 25 percent or so of all notarial site clauses ended up using the templates of landmark or vicinity. This exceptional gap is in no way a product of a misleading comparison between addresses and site clauses, for the addresses found in notarial acts used the templates of landmark or vicinity even *less* often, around 21 percent of the time.[42]

Specific examples make the general pattern clearer. The vicinity of the Spur was invariably called the "street of the Spur" in notarial site clauses from the mid-fourteenth century (5 usages), and sometimes the name was changed entirely, such as to "the street of Guilhem Folco." "Fruitery," used on fifteen occasions in the confraternal register, came out in notarial casebooks as "the street of the Fruitery" on three occasions and as "the Fruitery" only once. In a similar way, "Steps" was turned into "the street of the Steps" by the notaries, "near Guilhem Folco" ended up as "the street of Guilhem Folco," the market area known as "Tripery" became, in notarial hands, "the street of the Tripery," the "Palace" became "the street of the Palace," the "Shoemakery of the Temple" became "the street of the Shoemakery of the Temple," the "Arches" became "the street of the Arches," and so on.

We can see the translation process at work in the cases of specific individuals. The prominent buckler Johan Englese, a member of the confraternity of St. Jacques de Gallicia, gave his address to the scribe as the Jewish fountain. In 1359, in the midst of a kind of mini-territorial expansion, he purchased the two houses adjoining his property from the de Autu family. The notary who registered the transaction, Peire Giraut, listed the site as "the street of the Jewish fountain." In another case, the baker and confraternal member Guilhem Bidorlle, who identified himself as living in the Fruitery, was the host to the leasing of a vineyard drawn up in 1352. Johan Silvester, the notary who wrote the contract in Guilhem's house, identified the transaction site in the subscription clause as "the street of the Fruitery." In a third case, Rostahn de Mayron, who gave his address to the confraternal scribe as Frache, was, in a notarized house sale, named by a third person as owning a house located in "the street of the Frache." Last, the smith Esteve Bernat, who gave his own address as the "Spur," was identified in a notarized property sale as living in "the street of the Spur."

[42] See chapter 5, Table 5.2.

One could multiply these examples at length, and all would illustrate how notaries habitually translated vicinities and even landmarks into streets. The process carries the appearance of an innocent and relatively minor act of translation, made necessary perhaps by language difference but not otherwise meaningful. Yet I would argue against this complacent solution. One need not ascribe sinister motives to the notaries to appreciate that linguistic patterns reflect ideologies and power and that acts of translation invariably compromise meaning. In this case, notaries were translating units of sociability without fixed boundaries into immovable streets. Vicinities and landmarks, as cartographic entities, were eminently flexible, given that they could expand or contract according to the desire of people living reasonably nearby to be associated with them. Presumably the status of given vicinities rose and fell: there is no particular reason to assume that the Spur and the Jewish fountain—two morally sound vicinities of high status and desirability in the mid-fourteenth century—were equally desirable in the thirteenth century, or the fifteenth century. The flexibility of the construct allowed individual householders to identify themselves with the nearest high-status vicinity. In a street-based template, this flexibility is lost. The gradual ascendance of the notarial street-based model, to hypothesize on the basis of this evidence, eliminated one of the linguistic bases of social cartography and social identity in Marseille. This is not to say that social groups could not develop or continue to use unofficial descriptions of space. It is only to say that changes in linguistic cartography were making it discursively or cartographically more difficult.

In this light it is worth noting that Provençal, at least in the fourteenth century, sometimes won out. Twelve members of the confraternity described their address as the Jewish fountain, and never defined it as a street. In seven site clauses referring to the same area, the notaries called it the street of the Jewish fountain on three occasions but used the Provençal vicinity or landmark, Jewish fountain, on four occasions. Similarly, the Carpentery, which was used five times in the confraternal register and never called a street, was translated by notaries into a street only four times in nine site clauses referring to the area, even though the intramural Carpentery was so visibly a street in form. The historical process whereby Provençal cartography gradually gave way to notarial cartography was uneven, and what we have here is a snapshot taken in the mid-fourteenth century in which we find certain vicinities that even the notaries, who of course were Provençal speakers, did not always imagine as streets. Vicinities tended to remain vicinities, even in notarial usage, whenever they were seen as powerful markers of status

Table 4.3. Cartographic categories in vernacular addresses, by sex, 1349–53

	N	Streets (%)	Districts (%)	Landmarks (%)	Vicinities (%)
Women	84	10.7	7.1	27.4	54.8
Men	292	14.0	9.3	22.6	54.1

Source: ADBR 2HD E7.

and identity within the conventions established by Provençal linguistic cartography.

Female Cartography

The confraternal register of the hospital of St. Jacques de Gallicia is extraordinary because it reveals a Provençal and artisanal cartography unmediated by notaries and their Latinate norms. It is impossible to find a similar source for women unmediated by men, because there is no comparable source written by and about women—unfortunate, because there are good reasons, based in theories in cognitive linguistics, to think that women may have apprehended space in a way different from men. All the same, the register is one of our best sources for developing some kind of understanding of what might have been a distinct female cartography simply because so many women figure in it. In the register, however, we find that women were neither more nor less likely than men to use addresses in defining their identities: 62 percent of the men in the confraternity defined their identities by means of addresses, and so did the same percentage of women. The consistency is a little suspicious and probably indicates not that men and women were equally prone to self-addressing, but rather that confraternal scribes were proactively demanding addresses and were equally successful with men and women. A gender analysis of the addresses provided reveals that women were slightly less inclined than men to use streets and slightly more inclined to identify themselves by landmarks (Table 4.3).

However slight, the difference probably reflects the fact that women's interests were better served by the vicinal template; certainly a reputation for respectability mattered a great deal to a woman, and perhaps played a more important role in her access to resources than it did for men. Women may have stood to lose more in the gradual shift from vicinity to street.

CONTESTED SITES

One wonders, in fact, what Ugueta Provensala would have thought about her rather abrupt linguistic displacement, had she known of it. Ugueta was a fishmonger, and in August of 1380 the defendant in a criminal inquest. She had dumped foul water into the street, contrary to the city statutes. At some point during the preliminaries, Ugueta explained to the notary of the court that she lived in the Fishmongery, the fish market located in the *sixain* of Accoules. The notary duly transcribed her address. In the margin next to the address, however, another notary wrote, in slightly different ink and hence somewhat later in time, "in the alley or street of Peire de Serviers" (*in transvercia sive carreria Petri de Serveriis*).[43] It is a telling distinction. Ugueta the fishmonger thought of her living space in professional terms: it is the area where she and her mostly female colleagues sold fish and, with their husbands, made their residence. A court notary reviewing the transcript figuratively took her out of the vicinity and placed her in an alley. He probably did so because, in this part of the city, the street grid was relatively rectilineal. It was the alleys perpendicular to streets that ran down the gentle slope toward the port; hence, it was an alley down which the foul water had trickled.

This example illustrates well the argument of the preceding chapters that individuals from different linguistic communities were capable of imagining city space in distinct ways, and the distinction that exists in grammar between vicinities like the Fishmongery and streets or alleys like the street of Peire de Serviers captures this cartographic variance. Yet what we have in this particular case is not just a grammatical quarrel about the nature of the space—whether it is a vicinity or an alley—but also a lexical disagreement about the very name to assign to the area in question.

This issue surfaced in the section above because artisanal and Provençal vicinities were social constructs, not elements of the skeletal architecture of streets. If someone thought of herself as living within the Fishmongery, then that is what she gave as her address, regardless of whether officious notaries would locate her house in the alley of Peire de Serviers, or the street of Johan Sancho, or any other street that intersected with or overlapped the Fishmongery. It is not clear that Ugueta would have been willing to admit that she lived on a street named after a nobleman, although she must have been aware that some people used this name. As

---

[43] ADBR 3B 96, fol. 96r, 29 Aug. 1380.

a result, Provençal speakers and notaries each had their own lexicons. Because these lexicons overlapped to a large extent, the two linguistic communities agreed on the names to be given to most sites. The names of other sites, however, were contested. In the centuries before the reduction of maps to print and the standardization of street names under governmental authority, it was possible for any city to be filled with contested sites and a number of competing cartographic lexicons.

It is not difficult to find these contested sites, and they troubled Marseille's early topographic historians. "The name of the Cobblery was carried by the street of the Roiling Stone, the street of the Old Cobblery, and the street behind the Franciscans," complained Eugène Duprat in 1935. Duprat made the point to explain why the topographic history of medieval Marseille, at that time, was so little developed.[44] Bruno Roberty, operating in good, positivist fashion, smoothed over these distinctions in the map that he produced. Aware that the "street of the Brassworkery" and the "street of the Nettery" denoted the same street, as did Lancery and Panataria and a number of other names, he nonetheless implicitly suggested, in the index of names that accompanied his map, that one street name was to be preferred over others.[45] His predecessor in mapping, the nineteenth-century érudit J. A. B. Mortreuil, was somewhat more concerned; in a tiny hand he included all known street names on the onionskin maps he produced of the medieval city (Figure 8). The result is scarcely legible, with streets bearing, at times, four and five names. Hence, Mortreuil's maps suggest wild confusion in the topographic imagination of the residents of medieval Marseille. So too do Augustin Fabre's six volumes. In some respects Mortreuil's and Fabre's tacit suggestion of lexical confusion was correct, although the confusion was more apparent than real. In particular, the lexicons were relatively consistent within specific linguistic groups and in given generations. The apparent confusion found in Mortreuil's maps comes from the indiscriminate mixing of records, namely episcopal and notarial records kept in Latin and seigneurial records kept in Provençal, as well as records from different decades.

The existence of multiple cartographic lexicons made possible "lexical drift," the tendency, found in all medieval and early modern European cities, for street names to change over the course of centuries. One assumes that this process was not random, and there is an interesting history to be written of how and why some toponyms were effaced from

[44] Benoit et al., *Monographies communales*, 107.
[45] ADBR 22FF 86.

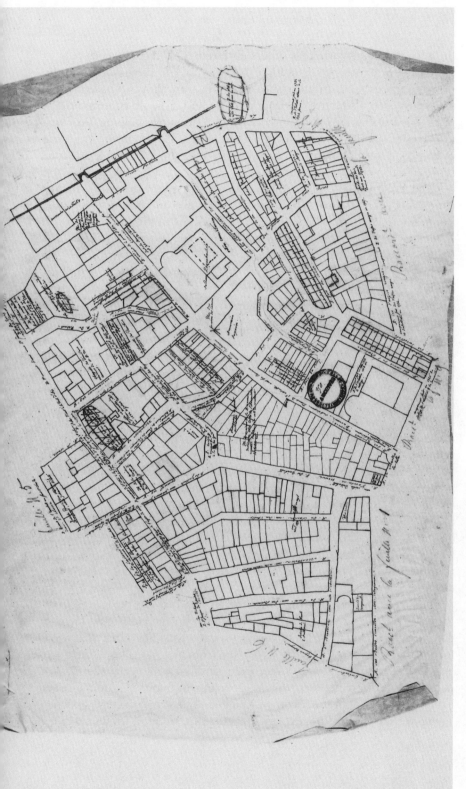

8. Portion of the Mortreuil map of medieval Marseille.

the sites of memory whereas others endured. This section seeks only to establish the existence of multiple lexicons in late medieval Marseille. Lords, notaries, and ordinary speakers of Provençal, as we have already seen, possessed distinctive spatial grammars, and it should come as no surprise to learn that they also used different, if often overlapping, lexicons.

The episcopal lexicon was the most distinctive. Shaped by the grammar of islands, it was not always prepared to use the name of a single street to represent the entire island and hence did not usually overlap with the notarial street lexicon. Instead, the episcopal chancery developed names commonly based on living residents, or their descendants, or the historical memory of them. There are some exceptions to this, as in the *insula Annonarie Superioris*, also the name of a street, but in most case episcopal islands were lexically unique.

The grammar of islands was a grammar of record-keeping, and did not interact with other templates and lexicons to any great extent. When streets replaced islands in the sixteenth century, they also eliminated island names, and city blocks came to be defined by their boundaries. Notarial and Provençal lexicons, however, jostled against one another in everyday usage and consequently shared a good deal both lexically and grammatically. The lexicons mutually interpenetrated one another, so that elements of the Provençal lexicon occasionally surfaced in notarial documents, and notarial lexical preferences were occasionally transliterated into Provençal. The differences, therefore, are intrinsically more interesting and were of two kinds. First, the Provençal vicinity, as a reflection of the social imagination of Provençal speakers, was not tied into the skeletal architecture of streets that was being developed within notarial usage, and hence the lexical terms did not necessarily translate easily from one template to the other. In some cases they did: the Carpentery, housed in a narrow intramural space, was identical with the street of the Carpenters. But many Provençal vicinities were agglomerations of what would be alleys and streets in notarial usage, and, accordingly, were routinely dissected by notaries. Sites bearing contested names reflect this fact.

The Provençal lexicon—or at least the lexicon that can be derived from the 376 addresses given in the register of the confraternity of St. Jacques de Gallicia—included 99 toponyms or lexical terms. The notarial lexicon for the mid-fourteenth century, as derived from 932 site clauses, included 225 lexical terms. Although the greater number of notarial site clauses makes comparison between the two lexicons difficult, it nonetheless is likely that the notarial lexicon was bigger than the Provençal lexicon. There are several possible explanations for this, none of which excludes the other. First, it may be the result of heavy recruiting by the confraternity from a few circumscribed areas: the bakers in the area of the Jewelery and the shoemakers on Negrel street, for example, seem to have joined the confraternity in considerable numbers. Second, the notarized site clauses found in Table 2.1 were taken from all property conveyances, not from residential properties only. These properties include gardens and workshops, and even houses could be used not as a living space but for storage. It is distinctly possible that certain areas of the city were more densely populated than others, and the smaller Provençal lexicon may reflect this uneven distribution. Third, certain vicinities carried high status and an individual living on or near such a vicinity was more likely to use it as part of his or her identity. Last, and most important, Provençal vicinities were bigger than streets, and in dissecting vicinities into streets, notaries necessarily multiplied the number of lexical items.

Notaries and confraternal members shared seventy-two lexical terms. These often appear as streets in notarial usage and as vicinities in Provençal—the "street of the Fruitery" and the "Fruitery"—but as we have seen, notaries could and did use the template of vicinity, and by the same token confraternal members could and did use streets. To anyone familiar with the map of medieval Marseille, these lexical items, whatever their form, stand out as familiar parts of the urban landscape. Use of these names by both social agents suggests that these were widely accepted and comparatively uncontroverted lexical terms.

Intrinsically more interesting are the twenty-seven items particular to the confraternal lexicon and the one hundred and fifty-three particular to the notarial lexicon. The items unique to the mid-fourteenth-century confraternal lexicon include such vicinities as Pretty Table, Auction, Labeurador, Lauret, Slipperery, Roiling Stone, and Shoemakery of St. Jacques. The list also includes a large number of landmarks, including five ovens not mentioned by the notaries and nine references to notable figures. There was only one street—the street of Guilhem Imbert—that was not used by notaries. The absence of these terms from notarial site descriptions does not necessarily mean that notaries were unfamiliar with them; they can show up elsewhere in notarial acts, notably in identity clauses. Perhaps houses located in these areas were not the subject of a conveyance between 1337 and 1362. Yet some simply had different names in the notarial lexicon. The Provençal vicinity known as Roiling Stone (*Peyra que Raja*) was known to the notaries as the street of the Cobblery. Pretty Table was roughly the same entity as a street known to the notaries as the street of St. Marthe. The Shoemakery of St. Jacques surfaces in the notarial lexicon as the street of Corregaria of St. Jacques, and the Slipperery was, to the notaries, indistinguishable from the street of the Coppersmithery or the plaza of the Pilings. The numerous references in the confraternal lexicon to notable figures and to certain landmarks, such as small ovens, was casual, and the notaries simply did not recognize these as elements of the increasingly official notarial lexicon. Only a few, such as the open spaces named after Jacme Cancel, Guilhem Folco, Raymon Rascas, and Peire Amiel, were shared by both. Strikingly, three of these four terms appeared as streets in the confraternal register. The exception was a space known to the notaries as the "street of Peire Amiel" that was listed in the confraternal register by means of the landmark template as "before Peire Amiel." This penetration by the street template into the confraternal register indicates that these three streets had, in a sense, been made official in notarial usage.

The notaries used 153 terms not used by the confraternity, and almost all were streets, alleys, and islands, although there were a few vicinities, such as the Panataria of St. Jean and the Coppersmithery. These terms include the names of a number of notable figures, but almost all have been translated into streets or alleys. The islands imported from the episcopal chancery that are found from time to time in notarial usage never appear in the confraternal lexicon. People living on them would have given their addresses to the confraternal scribe as Cavalhon or would have used the landmark template. What are listed in the confraternal register as relatively large and ungainly suburbs were sometimes, in notarial usage, divided into constituent streets, such as the street of Madam Capone, Gache street, the street of Madam Auriola, and the street of the Tanners, names never used by members of the confraternity.

Cases such as these show how inexact Provençal areas were carved into constituent streets, thereby multiplying the number of items in the notarial lexicon. There are a multitude of individual examples to illustrate the general rule. Johan Aymar, who gave his address as the Fishmongery around 1350, was placed by the notary Jacme Aycart in the street of Johan Sancho. From the records listed in the prosopographical index we know that Johan Sancho was a merchant who lived on a street located just below the Fishmongery. A woman named Covinens Raymbaud, who gave her address as the Quay, was located in 1359 on an alley running northward away from the quay named after the nobleman Laugier de Soliers. The brothers Bertomieu and Guilhem Estaca, both prominent shoemakers, gave their address as the Spur to the confraternal scribe, but several documents located their houses in the street of the Glaziery. The most interesting history may be that of the clothier, Antoni Gras, who gave his address to the confraternal scribe as the Drapery, and elsewhere gave his address as the street of the Upper Drapery, but in two notarial acts was identified by the notary Jacme Aycart as a resident of the street of the Almoner, a street that ran northward from the street of the Upper Drapery. The table below lists these and other examples of members of the confraternity whose addresses were changed in this way. In most cases, the notary used, as the address, a street within or intersecting with the named vicinity. Two entries provide examples of vicinities turned into islands in episcopal records.

This observation extends beyond the differences between confraternal and notarial lexicons, for even within notarial and seigneurial archives we find hints of contested sites, typically between address clauses and property site clauses. We have already seen how Ugueta Provensala, a self-described resident of the Fishmongery, was figuratively removed by a notary to the alley of Peire de Serviers in 1380. A similar case occurred a generation earlier, when Bertomieu Vincent's address was given by the notary, Jacme Aycart, as the street of the Fishmongery, but several years later Bertomieu showed up in a site clause as living on the street of Guilhem de Serviers, Guilhem being the ancestor of Peire de Serviers.[46] The house of Johan de Torreves, a self-described resident of the Jewish fountain, was located by a notary in "a dead-end alley next to the avenue of St. Martin" (*transversia que non transit iuxta carreria recta Sancti Martini*).[47] Exactly the same linguistic displacement happened to a buckler named Peire Gavot; a self-described resident of the Jewish fountain in one record, he too was relocated to this same alley in another.[48]

Vicinities popular in Provençal usage were confected out of adjoining and intersecting streets and, in notarial usage, were routinely cut up into constituent streets and alleys. People who identified themselves as living at the Hill (*Colla*) show up in the notarial lexicon as owning houses on the street of Perier and, some distance away, in the area of Negrel street. Within the confraternal register itself, a man named Nocho Ancona gave his address once as Negrel street and once as the Hill, perhaps indicating his own puzzlement about just what the area was called, unless he had just moved. Alternatively, the street segment around which the vicinity was organized was absorbed into a longer street. Consider the example of the Spur, which was given as an address by eighteen members of the confraternity. The vicinity, located near the northwest corner of the Jewry, just to the east of the church of St. Marthe, included not only what the

---

[46] ADBR 355E 9 131, fols. 105r–v, Jan. 1359, then 355E 11, fols. 33r–v, 30 July 1361.
[47] ADBR 355E 8, fols. 84r–v, 22 Feb. 1356.
[48] See ADBR 355E 11, fols. 158v–159r, 14 Mar. 1362; ADBR 351E 24, fol. 168r, 1361.

Table 4.4. Sites identified differently in vernacular and Latin cartography, 1337–62

Individual or proprietor	Address as defined in vernacular source[a]	Address or plat as defined in Latin source[b]	Type of clause[b]	Source[c]
Peire Austria	Jewish fountain	Street of Peire Austria	several	passim
Johan Aymar	Fishmongery	Street of Johan Sancho	address clause	355E 2, fols. 144r–v, 2 Jan. 1350
Raymon Bertran	Arches	Street of the Vivaut or street of the lord Berengier Vivaut	address clause	3B 45, fol. 147r, case opened 1 Aug. 1343
Nicolaua Capella	Corregaria	New street	site clause	4HD B1, fol 65v
Johan Escot	Glaziery	Island of Fulco Sardine	site clause	5G 114, fol. 187r
Bertomieu Estaca	Spur	Street of the Glaziery	site clause	358E 84, fol 119r–v, 2 Jan. 1352
Guilhem Estaca	Spur; also, before the oven of the street of the Almoner	Street of the Old Glaziery	site clause	381E 81, fol 14v, 7 Ma 1358
Guilhem Estaca	Spur; also, before the oven of the street of the Almoner	Street of the Old Glaziery	site clause	381E 82, fol 20v–32r, 2 Apr. 1359
Antoni Gras	Drapery	Street of the Almoner	address clause	355E 5, fol. 112v, 28 Jan. 1353
Antoni Gras	Drapery	Street of the Almoner	address clause	355E 6, fols 76r–78r, Sept. 135
Antoni Gras	Drapery	Street of the Upper Drapery	address clause	5G 114, fols 185v–186
Guilhem Jauselme	Cavalhon	Street of the Baths of Isnart Beroart	address clause	5G 144, fols 30v–31r
Guilhem Jauselme	Cavalhon	Island of Guilhem Sard	site clause	5G 114, fols 30v–31r
Guilhem Jauselme	Cavalhon	Street of the Baths near St. Cannat	site clause	351E 3, fol. 7v, 14 Jan 1354
Johan Jay	Gallican Gate	Street of Cavalhon, under the house of Isnart Beroart	site clause	355E 3, fols 3v–4v, 29 Mar. 135(
Rostahn Jay	Oven of the Cobblery	Street of Nicolau Grifen	subscription clause	355E 36, fol. 71v, 2 Aug. 135(
Nicolau de Lauzana	Slipperery; also, Pilings	Street of the Shoemakery of the Pilings	subscription clause	355E 6, fols 5r–v, 24 Apr. 135
Peire Minhoti	Fruitery	Street of St. Jacques	site clause	381E 82, fo 20v–32r, 2 Apr. 1359

ıdividual or ˙oprietor	Address as defined in vernacular source[a]	Address or plat as defined in Latin source[b]	Type of clause[b]	Source[c]
ıymon de Moysaco	Spur	Street of the Gavottas; also, street of the Bosquet	site clause	4HD B1 quatro, fols. 73r, 84v, 88v
ıfael Nicolai	Pilings	Street of the Nettery	site clause	355E 4, fols. 42r–43r, 22 Aug. 1351
ɔvinens Raymbauda	Quay	Alley of Laugier de Soliers	site clause	355E 9, fols. 99r–v, 10 Jan. 1359
ɖalays Renonta	Tripery	Street of the Tannery	site clause	381E 78, fols. 27v–28r, 13 Apr. 1350
ıurand de Verduno	Oven of Cavalhon	Cavalhon, in the island of Guilhem de Scala	site clause	5G 114, fol. 171v

All vernacular addresses in column two come from the register of the confraternity of St. Jacques de Galli-ɑ, ADBR 2HD E7.
ɕColumn 3 identifies either the address of the same individual according to a Latin address clause or the loca-ɔn of the individual's house according to a Latin site clause or subscription clause. The type of clause is ɕentified in column 4.
ɕColumn 5 provides the source for the Latin address or plat in column 3. Sources include notarial, seigneur-ɪ, and judicial records.

notaries called the street of the Spur but also several adjoining streets, including the street of the Gavottas, the street of the Glaziery, the street of Guilhem Folco, the street of the Almoner, and the Frache. These other streets occur much less commonly in the confraternal register.[49] The preference for using the Spur was evidently a result of the high status of the vicinity, an important center for stonemasons and bakers.

A particularly provocative possibility is that nobles and confraternal members possessed distinct lexicons. The Pilings or the plaza of the Pilings, located in the *sixain* of Accoules next to the port, was a place of work. So named because this was the only section of the shore that had pilings (*escars*), the plaza was directed toward the sea and maritime commerce. The maritime nature of the place was reflected by its alternate name, the *Scaria Navium,* or the Pilings of the Boats. The plaza was also associated with a shoemaker quarter, the Shoemakery of the Pilings. It was a favored element in the cartographic imagination of the confraternity of St. Jacques de Gallicia, used by twelve members to denote an address.

[49] The Glaziery appears once, the street of the Almoner three times, the street of Guilhem Fulco once, and Fracha twice.

At the same time, the plaza was associated with one of the most prominent noble families of Marseille, the Vivaut. In the noble lexicon, the
plaza was named after the family. A street called the street of the Vivaut
also ran eastward out of the square, further illustrating how the family
had inscribed itself on the city map.

Roberty identifies the Pilings and the plaza of the Vivaut as being the
same place. The names may have demarcated different sides of the same
plaza, a difference captured by the shoemaker Guilhem Faber, when he,
or the notary drawing up the act, gave his address in a 1354 land sale as
"the Shoemakery of the Pilings facing the plaza of the Vivaut."[50] Whatever the physical difference, if any, the members of the confraternity of
St. Jacques de Gallicia never described themselves as living on a plaza or
street named after the Vivaut, and the members of the Vivaut family never
described themselves as living at the Pilings. In documents from the mid-
fourteenth century Pilings was the more common name, and the plaza of
the Vivaut, when it occurs, is usually found in noble contexts—in notarial acts, say, involving a member of the nobility. Roberty favored "plaza of
the Vivaut," and did so because that name was still in use in his own day,
whereas Pilings had faded.

The names of nobles or noble families were routinely inscribed on the
landscape in this way. Examples are legion, including the plaza of the
Vivaut, the plaza of the Hugolin, the street of the Jerusalem, the court of
the Marquesas, the alley of Peire de Serviers, the baths of Isnart Beroart,
and any number of other streets, alleys, and islands. These names often
found their way into notarial usage, suitably translated into streets. Some
such names, like the street of Johan de Sant Jacme, a segment of the New
street, never became popular enough to acquire the stamp of notarial
authority. In artisanal usage areas named for nobles were much less
common. Artisans and commoners preferred thinking in terms of vicinities, and these were never named for noble families. When members of
the confraternity of St. Jacques de Gallicia, most of whom were artisans,
did use the names of notable individuals, the individuals mentioned were
either members of the confraternity, as in the case of the great merchant
and self-styled *nobilis* Peire Austria, or ordinary people, including prominent bucklers, masons, butchers, curriers, smiths, and merchants. Nobles
figured rarely in their lexicon. This may be because members of the confraternity simply did not live in areas named for nobles. It may also be
because they assigned different names to the same sites.

---

[50] ADBR 381E 80, fols. 97r–v, 18 Nov. 1354.

Thus, it is possible that there existed a considerable gap between common and noble lexicons, partly but not wholly mediated by the notaries. It is difficult to measure this gap with any precision because there is no document or set of documents that allows us easily to recover the noble lexicon. Noble preferences undoubtedly surface in notarial records, but without some independent standard it would be difficult for us to distinguish them. A few registers of rents owed to notable individuals exist in Provençal, probably written by the person to whom the rent was due. One of them, kept by the great merchant-nobleman Bernat Garnier, is characteristically Provençal in not using many streets, preferring to define areas as vicinities or in relation to landmarks.[51] It does not, however, use a notably distinct lexicon. A few areas were given unfamiliar names based on notable individuals, such as the street of Johan Elie and the street of Rostahn Bezenet, but these are the only two and hardly constitute a major departure from some lexical norm.

The Decline of the Artisanal and Retail Vicinity

Among the most interesting vicinities were those named after artisanal groups or retail trades. To judge by Roberty's index, there were, in all, around one hundred and seventy streets or alleys of any consequence in late medieval Marseille, and of this total forty bore names derived from artisanal or retail trades, appearing as vicinities or as streets, depending on the nature of the source.[52] Figure 9 shows the approximate locations of thirty-nine of these forty areas, and the accompanying legend gives the name according to the template of vicinity.[53] The location of the fortieth, the Bucklery, is unknown. Several, including the street of the Tanners (in the suburbs), the street of the Buttoners, the street of the Curriers, and the corner of the Oarmakers, always appear in the records in this way, that is, as streets or corners.

Clearly, certain artisanal and service trades were inscribed into the map

[51] ADBR 4HD B1 ter.

[52] These streets, of course, do not all have the same importance, for in the mid-fourteenth century a fair number, especially in the upper city, were lined with little more than gardens, pastures, or ruined houses, and had no permanent residents.

[53] For the sake of readability I shall call these areas "vicinities" and use the template of vicinity when referring to them, and it must be borne in mind that all these vicinities could also appear as streets in notarial practice.

9. Cartography of artisanal vicinities and streets, ca. 1350.

Latin name of neighborhood		English equivalent
1	*Agudaria*	Nettery
2	*Aurifabraria*	Goldsmithery
3	*Bladaria*	Mill
4	*Blancaria*	Tannery
5	*Carreria Blancorum*	Street of the Tanners
6	*Botaria*	Coopery
7	*Carreria Botoneriorum*	Street of the Buttoners
8	*Calafatia; Cantonum Magistrorum d'Aysie*	Caulkery
9	*Cambium*	Change
10	*Canabasseria*	Canvassery
11	*Candellaria*	Candlery
12	*Carreria Conreatorum*	Street of the Curriers
13	*Cordellaria*	Cordery
14	*Carreria Corderiorum*	Street of the Corders
15	*Cultellaria*	Cutlery

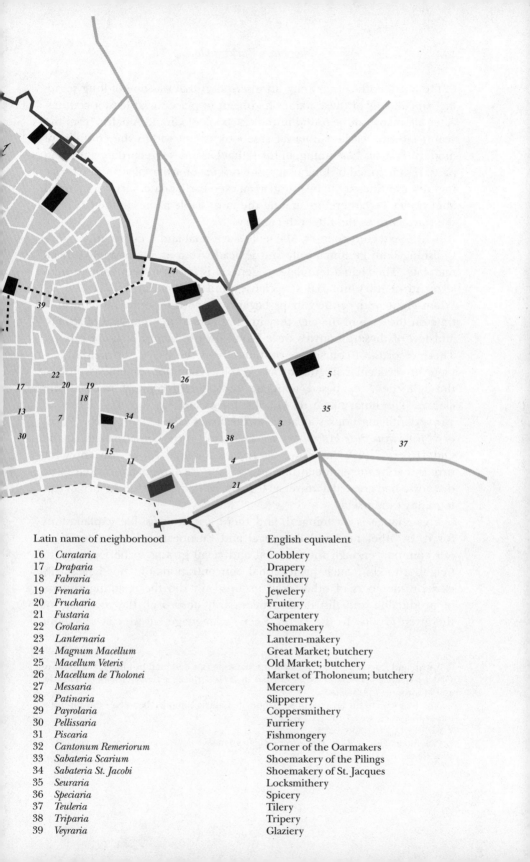

Latin name of neighborhood		English equivalent
16	*Curataria*	Cobblery
17	*Draparia*	Drapery
18	*Fabraria*	Smithery
19	*Frenaria*	Jewelery
20	*Frucharia*	Fruitery
21	*Fustaria*	Carpentery
22	*Grolaria*	Shoemakery
23	*Lanternaria*	Lantern-makery
24	*Magnum Macellum*	Great Market; butchery
25	*Macellum Veteris*	Old Market; butchery
26	*Macellum de Tholonei*	Market of Tholoneum; butchery
27	*Messaria*	Mercery
28	*Patinaria*	Slipperery
29	*Payrolaria*	Coppersmithery
30	*Pellissaria*	Furriery
31	*Piscaria*	Fishmongery
32	*Cantonum Remeriorum*	Corner of the Oarmakers
33	*Sabateria Scarium*	Shoemakery of the Pilings
34	*Sabateria St. Jacobi*	Shoemakery of St. Jacques
35	*Seuraria*	Locksmithery
36	*Speciaria*	Spicery
37	*Teuleria*	Tilery
38	*Triparia*	Tripery
39	*Veyraria*	Glaziery

of the fourteenth-century city, an inscription that was one of long standing, since many of these names had been in place for at least a century.
After all, many can be found in the casebook of Giraud Amalric from the
year 1248, the earliest notarial casebook extant within the confines of
modern France.[54] Assuming, in turn, that Giraud was recording a naming
pattern sanctioned by long usage, the origins of artisanal vicinities as cartographic entities are to be found at an even earlier date. Vicinities named
after trades were therefore among the most stable names in the city from
the thirteenth to the fifteenth century.

By the sixteenth century, Marseille's artisanal and retail vicinities were
vanishing both grammatically and lexically from an increasingly official
template. The vicinal template in general is found infrequently in mid-
sixteenth-century notarial site clauses and seigneurial rent registers.[55]
When eighteenth-century maps began to record toponyms for the first
time on the map of the city, they invariably used the template of streets,
and few of these toponyms were based on the names of trade groups.
These records, of course, say nothing about the use of the vicinal template in vernacular linguistic cartography. Vicinities sometimes pushed
through the street-based template in sixteenth-century notarial site
clauses. The notary Jean de Olliolis, for example, wrote site clauses in
1551 identifying houses located "at the street called 'The Narrow Furriery'" (*a la rue dicte la Pellisarie Estricte*) and "at the Street called 'At the
Old Tripery'" (*a la Rue dicte a la Triparie Vielha*).[56] All we can know for
sure is that by the eighteenth century, if not earlier, official cartographic
discourse had created a universal lexicon and grammar for the urban cartography of Marseille.

This change is meaningful, and there are two possible explanations
for it. In other medieval Provençal and European cities and towns, it
was common enough for artisanal and retail groups to be localized in
neighborhoods.[57] Such professional concentrations did not necessarily
exclude members of other craft groups, nor did the resulting centers
of production and distribution necessarily absorb all the members of
the given group. To use Derek Keene's language, all we can say about

[54] A map and legend of thirteenth-century Marseille can be found in John Pryor, *Business
Contracts of Medieval Provence: Selected "Notulae" from the Cartulary of Giraud Amalric of Marseilles,
1248* (Toronto, 1981), 64–65.
[55] A register kept by the hospital of St. Jacques de Gallicia compiled between 1588 and 1662
lists no vicinities; see ADBR 4HD B25.
[56] ADBR 373E 194, fols. 627v, 777r.
[57] Paul-Albert Février, *Le développement urbain en Provence de l'époque romaine à la fin du XIVe
siècle* (Paris, 1964), 165–67.

medieval Winchester is that trade groups had a "characteristic distribution."[58] The socioeconomic factors that encouraged localization, however, may have been lessening over the later middle ages, and artisanal communities in Provence may have been slowly disaggregating over the period.[59] The history of Marseille's trade groups and socioprofessional patterns of settlement, unfortunately, is markedly undeveloped, but there would be no particular reason to think that Marseille's trajectory in this regard would have differed markedly from the norm elsewhere in Provence. If artisanal toponymy reflects socioeconomic facts in some neat and uncomplicated way, then the disappearance of trade groups from the city map is a measure of socioeconomic change.

However plausible, this line of reasoning has some problems. The argument would work best if the inhabitants of a late medieval city constituted a single linguistic community and shared a common linguistic cartography. Yet, as we have seen, this was not at all the case. In Marseille we find several linguistic communities each with distinctive templates and possessing cartographic lexicons that did not entirely overlap. Thus, we cannot say that there was a single, universal map to which all subscribed and which reflected a consensus on matters of linguistic cartography. The decline of the artisanal and retail vicinity may reflect socioeconomic changes. But it may also reflect something quite different, the emergence of an official map shaped by the cartographic priorities of the emerging class of urban cartographers, in particular, public notaries and other agents of record.

Although I do not deny the possibility that linguistic cartography can reflect socioeconomic structures, in what follows I will pursue the argument that the gradual elimination of artisanal and retail vicinities reflects most clearly a process whereby the conventions of vernacular linguistic cartography were eliminated by an increasingly official and universalizing cartographic discourse. This may not mean that the physical decline of the artisanal and retail vicinity should be explained as the by-product of linguistic change. Instead, I think we should interpret both changes—

[58] Derek Keene, *Survey of Medieval Winchester*, 2 vols. (Oxford, 1985), 1:335.

[59] Evidence from the nearby cities of Arles and Aix also shows that by the fifteenth century, crafts were no longer closely associated with streets bearing their names (Noël Coulet, personal communication, Sept. 1992). See also Philippe Bernardi, "Métiers du bâtiment et techniques de construction à Aix-en-Provence à la fin de l'époque gothique (1400–1550)," thèse, Université de Provence Aix-Marseille I, 1990, 126; Michel Hébert, *Tarascon au XVe siècle: histoire d'une communauté urbaine provençale* (Aix, 1979): 65–69; Louis Stouff, "La population d'Arles au XVe siècle: composition socio-professionnelle, immigration, repartition topographique," in *Habiter la ville, XVe–XXe siècles*, ed. Maurice Garden and Yves Lequin (Lyon, 1985): 7–24.

linguistic and social—as the result of changes in the form and distribution of power.

To begin, it is very likely that artisanal and retail vicinities were real phenomena in twelfth-century cities and, moreover, that toponymy from the period accurately reflects the social and economic reality of these vicinities. With the sources extant from medieval Marseille, it is difficult to say exactly how or when the names of artisanal groups or retail centers were first inscribed onto the map of the city, or even in what form. The practice, however, is common enough in premodern cities. S. D. Goitein, for example, indicates that medieval Cairo had numerous streets and other areas of the city named after craft groups such as coppersmiths, turners, blacksmiths, alchemists.[60] André Gouron argues that artisanal names appear on the linguistic map of Languedocian cities from the twelfth century onward, although since he does not always distinguish carefully between streets and vicinities it is not easy to know which template was followed.[61] Since the vicinity was the preferred template of fourteenth-century speakers of Provençal, I think it is likely that artisanal and retail centers were originally known most commonly as vicinities, and that street-based versions of the name developed later, under the influence of public notaries and other agents of record. Ghislaine Fabre and Thierry Lochard cite two craft vicinities from Montpellier mentioned as early as the second quarter of the twelfth century—the Woolery (*Flocaria*) and the Tannery (*Blancaria*)—and a later vicinity, the Tintery (*Vermeilaria*), was first mentioned in 1183.[62] It is probably safe to assume that this toponymy originally recorded the social facts of artisanal production or residential patterns.[63] There are many economic reasons why we might expect to find thirteenth-century artisans or retailers clustered within vicinities in medieval cities, reasons having to do with the flow of materials, the control of prices, and hygiene.[64] Administrative surveillance is not least among these. Gouron cites a statute from 1204 forbidding members of certain professional groups from changing their

---

[60] S. D. Goitein, *A Mediterranean Society: The Jewish Community of the Arab World as Portrayed in the Documents of the Cairo Geniza,* 5 vols. (Berkeley, 1967), 1:83.

[61] André Gouron, *La réglementation des métiers en Languedoc au moyen âge* (Geneva, 1958), 115–39. Streets dominate in a footnote citing late fourteenth-century sources; see p. 127, n. 71.

[62] Ghislaine Fabre and Thierry Lochard, *Montpellier: la ville médiévale* (Paris, 1992), 65–66.

[63] In addition to Gouron, see Jean-Pierre Leguay, *La rue au moyen âge* (Rennes, 1984), 130–33.

[64] Marseille's statutes themselves stipulated that certain crafts, such as caulkers, fishmongers, and tanners had to practice their trades in areas where waste could be easily drained.

places of residence.[65] These neighborhoods, of course, did not have to form this way. Ronald Weissman describes a Florence that had a multitude of relatively self-sufficient neighborhoods with a full spectrum of crafts.[66]

The cartography of artisanal communities may also have reflected a political order. As noted in the introduction, very early in the thirteenth century Marseille developed a political order in which representation was based on guild membership. Artisanal vicinities probably played a role in this political order. We can assume that anyone aspiring to leadership in a given craft and hence to any sort of power in the city needed to reside in the appropriate neighborhood for two reasons: first, in order to be in constant contact with potential electors; second, in order to prove good faith. Lesser members of a given craft may have profited from living in an artisanal vicinity in order to receive the favors that masters, as politically powerful men in the early thirteenth century, were able to dole out. The interests of power can be phrased yet another way: masters may have promoted common working or living spaces the better to supervise the actions of journeymen and apprentices.

Whatever the historical reality of artisanal vicinities, there is no particular reason to think that fourteenth-century toponymy continued to reflect the social facts of artisanal and retail organization in some neat and uncomplicated way. Among other things, the forty named areas in no way represent all the trades practiced in Marseille, for conspicuously absent are clothiers, weavers, painters, fullers, fletchers, hosiers, parchment-makers, and a host of additional trades of lesser importance, let alone low-status professions like fishermen, mariners, and laborers. Furthermore, such communities can be mobile, and the social or economic environment that might encourage close domiciliation or common working quarters in a given era can change, leading to the dissolution of artisanal communities as social facts. Even if cartographies record the sites of memory, the relationship between memories and facts is something that must be explored, not taken for granted.

In Marseille, it is possible to isolate several historical processes that con-

---

See, for example, Marc Dupanloup, "La corporation des cuiratiers à Marseille dans la première moitié du XIVe siècle," *Provence historique* 77 (1969): 189–213. The statutes are presented and discussed in Pryor, *Business Contracts*, 82–84. For these three crafts, see also *Statuts*, pp. 147, 187; p. 59; and pp. 51, 113.

[65] Gouron, *Réglementation*, 127.

[66] Ronald F. E. Weissman, *Ritual Brotherhood in Renaissance Florence* (New York, 1982), 9–10.

tributed to a growing gap between the realities of socioprofessional settlement patterns on the one hand and urban toponymy on the other. All were related to the notarial tendency to see the city as a tracery of streets and to the growing authority of notarial cartography. None is mutually exclusive.

I begin with practices of dissection. As seen above, notaries had a tendency to dissect landmark vicinities into an array of adjoining or overlapping streets, each with a different name. A similar tendency operated in the case of artisanal and retail vicinities. Consider the caulkers. The caulkers, known in Latin as *calafati* or *magistri d'aysie*, were linguistically associated with a district known as the "Caulkery" (*Calafatia*) or, more often, the "corner of the Caulkers" (*cantonum Magistrorum d'Aysie*), located in the *sixain* of St. Jean near the port. The trade was an important one to a port like Marseille and there were a lot of caulkers in the city. Of their number the prospographical index for 1337 to 1362 provides a good idea of the domiciles of twenty. Only one of these can actually be identified as owning a house within the named vicinity. This was Johan Barratan, who, while giving testimony in a tangled inheritance dispute heard in court in 1339, identified himself to the judge as a caulker living "in the Caulkery, near the Coopery."[67] Others lived scattered across the city, several in the intramural Tannery. What we seem to have here, then, is a case in which the cartographic lexicon recorded a memory, not a social fact. But the pathetic figure of one of twenty is deceptive. Seven of these men can be identified as living on streets located in the *sixain* of St. Jean quite close to, and in many cases overlapping with, the corner of the Caulkery, such as the street of Lancery, the street of the Panataria of St. Jean, and the street of the Coopery. Provençal sources lacking, we cannot be sure that the men living on these streets would have identified their living space as the Caulkery, but the very flexibility of the vicinal template, the ability of a vicinity to stretch across several streets and alleys, suggests strongly that they would have. Dissected into streets by notarial cartography, however, the Caulkery seems to vanish as a social fact.

The process was at work elsewhere. The prosopographical index identifies no less than twenty shoemakers who lived in or near the Pilings. Only seven can be found specifically in the Shoemakery of the Pilings; the others were located, in a variety of documents, in the street of the Nettery, in the street of the Pilings, in the Pilings, in the Pilings near the street of the Nettery, and in the Slipperery. Tellingly, these seven were men who had

<hr />

[67] ADBR 3B 38, fol. 65v, 14 Oct. 1339.

the opportunity to define their place of residence in the register of the confraternity of St. Jacques de Gallicia. Four smiths were located just above the street of the Smiths on the New street and on the street of the Jewelery. In their cartography, the "Smithery" may have existed at this intersection. One of them, Bertran de Cruce, made just such an equivalence. In a site clause his house was located on the New street, but Bertran gave his own residence in another document as the Smithery.[68] Of the ten corders whose domiciles we know, none lived specifically in the two vicinities named for the trade, namely the Cordery, located in the *sixain* of Draparia, and the street of the Corders, located in the suburbs just outside the gate of the Frache. Six, however, can be identified as living on nearby or overlapping streets or areas, such as Frache and the suburb of Prat d'Auquier. In this case the topography was complicated by the fact that the corders, who presumably at one time practiced their trade in the intramural Cordery, were translating themselves to the street of the Corders in the suburbs. One continued to live and work in or near the old Cordery, and five had moved near the new street.

The Candlery was named by one candler, Jacme Tibaut, who identified his place of residence in a quittance as the Candlery of Corregaria.[69] The houses of two other candlers can be located in the general area known as Corregaria but not specifically within a vicinity called the Candlery.[70] Thus, the three men may all have been living nearby one another in an artisanal vicinity, but this probable social fact is hidden from us. Clearly, the name Candlery did not circulate widely outside candler circles. Indeed, there is another possibility: Jacme Tibaut was the *only* person in any extant record from the mid-fourteenth century to have called a part of the city the Candlery. We cannot be sure that other candlers would have agreed with him; several, in fact, lived near the Fishmongery, and another group lived near the church of Notre Dame des Accoules. The artisanal habit of identifying a vicinity was itself a discursive structure, not a social fact.

Let us consider a last example, that of the curriers (*conreatores pellium*). Marseille was a producer of leather products and the numerous curriers of Marseille handled an important stage in the treatment of hides. Houses belonging to thirteen can be identified in the prosopographical index, and all were located near the street of the Curriers south and west of Negrel street. Clearly the curriers were geographically organized into an

---

[68] ADBR 3B 48, fol. 82r, case opened 5 Oct. 1351; ADBR 6G 485, fol. 8v.
[69] ADBR 355E 35, fol. 35v, 19 May 1357.
[70] ADBR 355E 34, fols. 57v–58r, 16 Feb. 1350; ADBR 5G 115, fol. 7r.

artisanal quarter that was as coherent as, say, the Goldsmithery, even though the name of the area did not have the characteristic grammatical structure of a vicinity. Only one of these men, however, can actually be located on the street of the Curriers itself; others owned houses on the nearby streets of Negrel, Colombier, Almoner, and Castilhon.[71] There is no evidence suggesting that curriers, like tanners or carpenters, worked in workshops that were distinct from their domiciles. A distinction between workplace and residence, in other words, cannot explain this gap. The singular problem for curriers is that their craft was cartographically attached to a street, the street of the Curriers, and not to the linguistically more flexible vicinity. A vicinity is a geographically vague designation that can stretch to encompass several streets. It is not clear why the entire area had not come to be known as the "Curriery" (*Conreatoria*) or some similar name, given the dominance of the curriers in the vicinity.

These examples could be multiplied, and all would reveal the possibility of a significant gap between the cartographic imagination of notaries and that of artisanal and retail groups. Notarial cartography could not and did not acknowledge adequately the social facts of artisanal distribution and organization, social facts that were reflected better by the flexible grammar of the artisanal vicinity.

Dissection was one process that contributed to the decline of the artisanal vicinity as a feature of the linguistic cartography of the city. There was also a second process at work, for the standard cartographic lexicon was becoming increasingly resistant to the introduction of new names based on artisanal groups. According to the prosopographical index, there were several artisanal groups in the mid-fourteenth century whose members show a clear preference for living and probably working close by one another (Table 4.5). The stonemasons, centered in the Spur (six individuals) and around the Jewish fountain (seven individuals), the painters in the Carpentery, the clothiers in the Drapery, and the shoemakers on Negrel street comprise the most prominent artisanal groups to have settled in economically real but cartographically unmapped vicini-

---

[71] The sole currier who can be located on the *carreria Conreatorum*, Bertran Scutifer, identified the street as his place of residence while making an acknowledgment of episcopal lordship over a vineyard he owned in the hinterland. The document was dated 1353. Curiously, twelve years later, in the course of making a similar acknowledgment, Bertran apparently gave his address as an alley, namely the *transversia de carreria Negrelli*. The two labels probably identified the same street. See ADBR 5G 114, fols. 114r–v, and ADBR 5G 116, fol. 33r.

Table 4.5. Social topography of artisans belonging to trade groups without formal vicinities, 1337–62

	N^a	Lived near colleagues in *de facto* artisanal vicinity[b] (%)	Lived apart from colleagues (%)
Clothiers	28	67.9	32.1
Stonemasons[c]	21	61.9	38.1
Painters	10	70.0	30.0
Hosiers	7	71.4	28.6
Fletchers	4	100	0
Fullers	4	75.0	25.0
Parchment-makers	3	66.7	33.3
Total	77	68.8	31.2

Source: Prosopographical index.
[a] Column 2 provides the number of tradesmen belonging to a given trade whose addresses are known.
[b] Column 3 identifies the percentage of members of the given trade group who lived on the same or adjoining streets.
[c] The stonemasons formed two vicinities.

ties, and their failure to have their names inscribed on the map is somewhat surprising. This failure cannot be explained simply as a result of political insignificance, for groups such as the lanternmakers, tilers, and candlers were politically insignificant, and yet their artisanal identities were mapped onto the city.

An explanation for this cartographic oversight lies in the public nature of the areas in which they settled. These crafts were located in areas heavily interpenetrated by other crafts and other status groups, such as nobles and merchants. In the case of the stonemasons, the Spur, as we have seen in the previous section, had its own identity to which its residents were attached, and this seems to have undermined the possibility of naming the vicinity the Masonry (*Peyraria*) or some such name. In the case of the clothiers, their trade was swamped by the greater prominence of the drapers among whom they worked, and the same is true for the painters in the Carpentery. Negrel street was very long and included many other trade groups, sixteen in all; in addition, it was a major north-south thoroughfare and hence had a very public identity.

But this is not the only possible explanation for the lack of cartographic recognition for these trade groups, for it is distinctly possible that these areas had been settled only recently by members of the trades, and that names simply had not developed. Artisanal movement was common enough elsewhere in late medieval European cities, although a careful study of socioprofessional patterns in thirteenth-century Marseille would be necessary to prove the point. If this is correct, the failure of newly implanted artisanal groups to place their stamp on the cartographic lexicon of the city may be a result of the political insignificance of artisanal groups. Following the Treaty of the Peace of 1257, craft groups were denied a direct voice in governing, presumably because Charles of Anjou recognized that powerful resistance to his claims came from craft groups. The rise of an oligarchic order in Marseille dates to this period, the second half of the thirteenth century.[72] As a result of this political transformation, there were profound changes in the distribution of power and resources, changes that promoted nobles and wealthy merchant-entrepreneurs at the expense of artisans. It is not uncommon to find fourteenth-century streets, let alone islands, renamed simply because a nobleman or great merchant lived there. The street segment named after the nobleman and merchant Peire Austria, a fairly recent immigrant from Montpellier, is a case in point. This was not happening with artisanal groups.

In fact, very much the opposite was taking place. The intramural Tannery provides an especially significant example, for although the name remained, it was no longer home to any tanners, who for reasons having to do with odor and sanitation had moved outside the city walls some time before.[73] In the mid-fourteenth century the Tannery had been settled by laborers and a wide array of other trades or status groups, most notably nobles and merchants. Gradually the old name was lost. A register of rents owed to the crown from 1377 records a stage of this process. The register, organized by island, named one after the nobleman Johan de Cuges.[74] But the island's name had evidently changed, and since this was a potential point of confusion the notary saw fit to mention the previous name of the island, the Island of the Lower Tannery. This reveals a drift in the lexical term away from artisanal groups and toward noble patrons, a drift that reflects, in a small way, the political defeat of craft guilds in the thirteenth century and the gradual process whereby artisanal groups were expunged from the sites of memory and cartography.

Drawing together the threads of this section, I would like to offer a

---

[72] See Georges Lesage, *Marseille angevine* (Paris, 1950), 166, and Christian Maurel, "Le prince et la cité: Marseille et ses rois . . . de Naples (fin XIIIe–fin XIVe siècles)," in *Marseille et ses rois de Naples: la diagonale angevine, 1265–1382* (Aix-en-Provence, 1988), 91–98.
[73] Dupanloup, "Corporation."
[74] ADBR B 831, fol. 47r.

possible explanation for the processes that lay behind the linguistic elimination of artisanal vicinities that gradually took place across the later middle ages and was substantially complete by the sixteenth century. Before the emergence of the public notariate there were few professional cartographers outside seigneurial circles, that is to say, few agents of record whose activities included the act of recording of the landscape in some legal or official way. Let us assume that toponyms developed according to social logic and were established as vernacular linguistic conventions. They were rarely written down and were not reified by the act of writing. The ability of artisanal vicinities in the fourteenth century to stretch over an indeterminate area reveals both the portability of these names and, in vernacular circles, their social or cartographic utility.

The public notaries, whose presence in Marseille and elsewhere grew exponentially from the thirteenth century onward, and who quickly outstripped the cartographic authority of seigneurial curias, had a different cartographic agenda. The legal nature of their acts demanded a certain degree of precision and, hence, the notarial gaze tended to petrify the landscape. In the case of landmarks such as churches, which were not portable, the fixing of the street name did no great violence to vernacular cartographic conventions. In the case of artisanal vicinities, however, the fixing of names divorced the toponym from the craft.[75] A gap emerged between the word and the thing it originally signified, and the gap widened whenever artisans, in the normal course of things, changed their place of domicile and professional activity.

This process of petrification was enhanced by the notarial tendency to translate vicinities into streets. A particularly striking feature of the moves of both the tanners and the corders is that, in both cases, the original artisanal vicinities were known as free-standing nouns, the Tannery and the Cordery, whereas the newly settled areas were both known as streets, the street of the Tanners and the street of the Corders. This indicates, as argued earlier, that the template of vicinity was the template most commonly used in the twelfth or thirteenth century, and artisanal areas newly settled in the fourteenth century were named according to the increasingly hegemonic street template. At the same time, existing vicinities were being translated into streets. As was the case with the curriers and the street of the Curriers, streets are not flexible and cannot stand in for

[75] A systematic reading of Adrien Blès, *Dictionnaire historique des rues de Marseille* (Marseille, 1989), shows more generally how few medieval names remained in existence in the early modern era, five to ten at most.

vicinities as social constructs. They are tied to the skeletal architecture of streets, not into the social function of craft groups. They remained in place whenever artisanal groups moved on, and became increasingly moribund.

There was another process under way that affected all vicinities, artisanal and landmark alike. Over the course of several centuries, notarial cartography progressively lengthened some streets, eliminating in the process the street segments that had once formed the linguistic roots for vicinal identities. Emblematic of this tendency is the remarkable case of the New street. Running from the gate of Tholoneum to the street of the Upper Drapery, it was one of the longest streets of the lower city. But it was truly a new street, at least in name, and perhaps for that reason was not often used by speakers of Provençal. In the register of the confraternity of St. Jacques de Gallicia, for example, men and women never referred to the New street; instead, they cut the street into its constituent segments, using names based on Provençal usage such as Jewish fountain, Jewelery, Fruitery, and Tholoneum. The notarial use of the New street absorbed these vicinities into the larger entity. The street of the Lesser Drapery was a similar construct, because it enfolded two vicinities known in Provençal as the Spicery and Corregaria; so too was the street of the Lancery, which similarly enfolded areas known as Nettery, Coopery, Brassery, Slipperery, and Panataria. In preferring these long streets over vicinities, notaries were anticipating Baron Haussmann by some five hundred years in creating ever longer streets, bulldozing their way, in linguistic fashion, through Provençal vicinities as they did so.

This trend has an interesting parallel in the architectural re-engineering of streets found in some Italian cities in the fifteenth century.[76] As Charles Burroughs has observed, this re-engineering was associated with a growing communal interest in promoting wider and straighter streets as a strategy for eliminating noble encroachments.[77] In Marseille, perhaps owing to the failure of communal ideology and the political dominance of an oligarchy comprised of nobles and great merchants, there were few architectural transformations before the sixteenth century. Here, the re-

[76] E.g., Jacques Heers, *Espaces publics, espaces privés dans la ville: le liber terminorum de Bologne (1294)* (Paris, 1984), 129–32.

[77] Charles Burroughs, "Spaces of Arbitration," forthcoming in *Medieval Practices of Space*, ed. Barbara Hanawalt and Michal Kobialka (Minneapolis, 1999). See also Paula Lois Spilner, *"Ut Civitas Amplietur": Studies in Florentine Urban Development, 1282–1400*, Ph.D. diss., Columbia University, 1987.

engineering of streets was linguistic before it was physical, and the chief agent of the linguistic re-engineering was the city's public notariate. The arrival of the French language and French royal administration, starting in the late fifteenth century, probably pushed this process along. Other medieval street names based on ancient Provençal originals would survive into the twentieth century; vicinities and streets named for crafts were the first to go.

Did this change in language have any significant influence on practice? George Lakoff has recently argued that the structure of language can indeed influence nonlinguistic behavior, although not necessarily for the reasons posited by Benjamin Lee Whorf.[78] As Lakoff and Mark Johnson suggested in an earlier work, language is suffused with metaphors, the use of which allows complex ideas to be conveyed in terms of simpler cognitive experiences.

> Many of our activities (arguing, solving problems, budgeting time, etc.) are metaphorical in nature. The metaphorical concepts that characterize those activities structure our present reality. New metaphors have the power to create a new reality. This can begin to happen when we start to comprehend our experience in terms of a metaphor, and it becomes a deeper reality when we begin to act in terms of it. If a new metaphor enters the conceptual system that we base our actions on, it will alter that conceptual system and the perceptions and actions that the system gives rise to. Much of cultural change arises from the introduction of new metaphorical concepts and the loss of old ones.[79]

In the case of linguistic cartographies, we are dealing with categories, not metaphors, but the principle is much the same. As notarial cartography became the official cartography in the seventeenth and eighteenth centuries, a single category of official spatial apprehension, the street, was promoted at the expense of vicinities and landmarks. This development in linguistic cartography has a parallel in the growth and spread of a standard national language and was just as much related to the nation-building project.

To conclude, let us consider trends in language and identity in the

---

[78] George Lakoff, *Women, Fire, and Dangerous Things: What Categories Reveal about the Mind* (Chicago, 1987), 330–37.
[79] Lakoff and Johnson, *Metaphors*, 145.

broad perspective. Charlemagne, like his predecessors, styled himself the "King of the Franks," that is to say, the leader of a people, a *gens*. So did the earlier Capetian rulers. The ensuing centuries saw a remarkable transformation, as the dominion of kings became a region and not a race, a transformation reflected in the new preference for the title "king of the French," and eventually the "king of France." France, as a geographical entity, had replaced the Franks as a people.[80] Compare this to the trends that characterized the mapping of artisanal groups in medieval Marseille. At some point during its commercial rebirth in the eleventh and twelfth centuries, for example, the city of Marseille acquired a Tannery. This was a geographical entity, like "France" but on a much smaller scale. The Tannery probably once identified a space where tanners lived and worked. It was both a place and, if you will, a people. Over the course of the thirteenth or fourteenth century the Tannery lost its association with tanners. Although the name continued to be part of the cartographic imagination throughout the fourteenth and fifteenth centuries, it was falling into desuetude and would vanish by the sixteenth or seventeenth centuries. In its place there arose, for a time, a street of the Tanners. This street was located in the suburbs. Newly settled, it never acquired a toponym (the Tannery); it was reduced in status to a street where tanners happened to work. This street, too, was gradually eliminated from the cartographic lexicon in the sixteenth or seventeenth centuries. By the eighteenth or nineteenth century whatever was left of the tanning trade in Marseille would have become a trade—a people, that is to say, bound together by their labor but no longer inscribed in the city's cartography, no longer a living part of the city map.

The two trends moved in opposite directions, in the one case from a people to a place, in the other from a place to a people. It is not too far-fetched to suggest that the trends were related so that the invention of a national and mappable identity necessarily eliminated crafts as spatial entities. As we have seen in this chapter, artisanal and landmark vicinities played a large role in the cartographic imagination of artisans, retailers, professionals, and laborers. These vicinities denoted spaces with imprecise boundaries defined more by their social and economic function than by their geographical location. The use of landmarks was also preferred by everyday speakers of Provençal; landmarks, too, denoted imprecise and somewhat unmappable areas. Both usages appear in notarial records,

---

[80] See Colette Beaune, *The Birth of an Ideology: Myths and Symbols of Nation in Late-Medieval France*, trans. Susan Ross Huston, ed. Fredric L. Cheyette (Berkeley, 1991).

but both were declining over the course of the later middle ages as notaries promoted the use of streets. The emergence of notaries as cartographers helped petrify the urban cartography of Marseille. Notaries did so not by freezing the cartographic lexicon; names of open spaces would continue to evolve, and do so to this day, although now the process is under the conscious direction of institutions of the state. Notaries in medieval Marseille petrified the landscape by replacing cartographically imprecise although socially and economically meaningful vicinities and landmarks with a skeletal architecture of streets increasingly stripped of their social meaning. The names attached to the streets, in the process, ceased to signify. As we shall see in the last chapter, this transformation in cartographic imagination laid the groundwork for the fundamental remapping of personal identities that has become so marked a feature of modernity.

CHAPTER FIVE

Identity and Address

Ll records, by virtue of what they are, identify people and things. Strategies of identification, however, vary from one record-keeping culture to another. As a general rule, these strategies fall on a continuum stretching from social memory, at one extreme, to the depersonalized identity clauses of modern record-keeping bureaucracies, on the other. In many record-keeping bureaucracies throughout medieval Europe, the act of identification was typically a function of memory, and as a result, fiscal documents, judicial records, private acts, and other records that sought to identify people often included nothing more than a name.

Consider the following identity clauses, drawn from a record kept by the *clavaire* of Marseille in 1407, recording the receipt of fines levied by the court of inquest for criminal activity:

> Ysabella, a fallen woman
> Symonet Drapier
> Argentina, wife of Symonet
> Picardello
> Johan Le Bus, baker of Marseille.[1]

Compare this to a list of people identified in a record of criminal condemnations from 1907, exactly five hundred years later, in which personal identities look like this:

---

[1] ADBR B 1943, fol. 6r.

[188]

Serny, Agnès Célerine Joséphine, 32 years old, teacher, born in Roquefeuil
(Aude), residing in Marseille in the street of St. Gilles, no. 10

Castellotti, Joseph Louis, 18 years old, seaman, born in Bastia (Corsica),
residing in Marseille in the street of Figuier de Cassis, no. 8

Amato, Giovanni, called Paniotti, Charles, 34 years old, cobbler, born in
Viltoria (Italy), without domicile

Biu Van Dông, 18 years old, *rotinier*, born in Nhi-Hua (French
Cochinchina), residing in Marseille in the street of the Holy Family

Peyron, Berthe Jeanne Albine Joséphine, 28 years old, born in Marseille
(Bouches-du-Rhône), living there in the avenue of the Prado, no. 68.[2]

Much has changed in five hundred years. Both records use names, but
the common ground ends there. There was no template in 1407. If we
have learned that Johan Le Bus was a baker, it is surely because Johan
just happened to mention this fact to the official who kept the record. In
contrast, by 1907 the template is preprinted on the form: name, age, profession, birthplace, and—most important as far as this book is concerned—address, or place of domicile. One can imagine the questions
rattled out as the functionary consulted the form in front of him, and it
is not difficult to imagine the conclusions the police would have drawn
from the address given, or rather not given, by Giovanni Amato. This does
not necessarily mean that the agent of the police who filled out the record
did not know Giovanni personally. It only means that the identity he constructed had to identify a unique individual regardless of who consulted
the record. His own putative personal knowledge of Giovanni, in principle, was stripped from the record.

As discussed in the introduction, the official place of domicile has
become one of the most important elements of modern bureaucratic
identity templates because, like a name, it helps to identify a unique individual. The absence of any addresses in 1407 tells us that the very idea of
an address is a historical invention. Address, in fact, is not a natural form
of identity, because people often move. Societies that practice swidden
agriculture or nomadic pastoralism may have an idea of home—they certainly have a concept of hospitality—but the home in question is not fixed
and hence not easily mappable. The idea of a fixed home seems more
natural in urban or settled agricultural societies, but even here questions
of life cycle and gender intrude. In many societies, such as those of
medieval Europe and the Mediterranean, young men do not necessarily
have a fixed abode and even older men may absent themselves from the

[2] ADBR 403U 635 (Jan. 1907), dossiers 12, 59, 66, 102, and 124.

home for extended periods of time.[3] As S. D. Goitein has remarked, "Mediterranean man in the Middle Ages was an impassioned and persevering traveler."[4] The saga literature of Iceland is nothing if not a litany of the restless movements of homesteaders over land and sea.[5] Both women and men can be highly mobile in their marrying years. Work-related immigration was exceedingly common in the middle ages; on the other end of the social spectrum, aristocratic families could be highly mobile.[6] Geographic mobility, in sum, was far more prevalent in the pre-modern world than nineteenth-century social theory imagined. In the middle ages, Icelandic chieftains, French counts, and Florentine and Sienese patricians often associated themselves with houses, but to us these houses do not feel like addresses; they do not have the same cellular format and architectonic form that the address has, and they are not identical with a legal domicile.[7]

Given this state of affairs, the very idea of a permanent address or legal domicile, wherever it is found, is an artificial construct. It is not the same thing as the universal concepts of home and hospitality. Part of the impetus for defining domicile comes from states and their bureaucracies, which have some interest in making a link between individuals and place of residence. The domicile as a legal aspect of identity must be as old as the concept of a citizenry itself, even if what is being described here is residence within some geographical boundary, not the precise street and number we associate with the address today.[8] More pertinent, the

---

[3] Arguments concerning *iuvenes* are made by Georges Duby in *The Chivalrous Society*, trans. Cynthia Postan (Berkeley, 1977), 112–22.

[4] S. D. Goitein, *A Mediterranean Society: The Jewish Community of the Arab World as Portrayed in the Documents of the Cairo Geniza*, 5 vols. (Berkeley, 1967), 1:273. See also 1:42–59.

[5] For some discussion of this literature, see William Ian Miller, *Bloodtaking and Peacemaking: Feud, Law, and Society in Saga Iceland* (Chicago, 1990), 111–37, esp. 123 (the circulation of children between families), and 135 (the mobility of men by reason of famine, feud, and overpopulation).

[6] Important works include J. Ambrose Raftis, *Tenure and Mobility: Studies in the Social History of the Mediaeval English Village* (Toronto, 1964); L. R. Poos, *A Rural Society after the Black Death: Essex 1350–1525* (Cambridge, 1991), 159–79; Johan Plesner, *L'émigration de la campagne à la ville libre de Florence au XIIIᵉ siècle*, trans. F. Gleizal (Copenhagen, 1934); Robert Bartlett, *The Making of Europe: Conquest, Colonization and Cultural Change, 950–1350* (Princeton, 1993), 24–59.

[7] Some of the important works on this question include Karl Schmid, *Gebetsgedenken und adliges Selbstverständnis im Mittelalter* (Sigmaringen, 1983); Duby, *Chivalrous Society*, 86–87; Christiane Klapisch-Zuber, *Women, Family, and Ritual in Renaissance Italy*, trans. Lydia G. Cochrane (Chicago, 1985), 78–80, 117–18; Samuel K. Cohn, Jr., *Death and Property in Siena, 1205–1800* (Baltimore, 1988), 152.

[8] In certain legal contexts, for example, what mattered was not the street address, but rather the fact of residency *grosso modo* within a given political jurisdiction. On this issue, see Julius Kirshner, "*Civitas Sibi Faciat Civem*: Bartolus of Sassoferrato's Doctrine on the Making of a

demands of state finance, in particular, property taxes, poll taxes, and census, required a careful understanding of exact place of domicile. In Roman Egypt and probably elsewhere in the Empire, citizens were required to return to a legal or fiscal domicile during the taking of the census.[9] A similar situation obtained centuries later in the caliphate of Egypt.[10] In Ch'ing dynasty China and Tokugawa Japan, states exercised close control over people through neighborhood associations and registration; both the Chinese *pao-chia* and the Japanese *gonin-gumi* systems were based on an enumeration of households, the numerical rationality that seems so marked a feature of rational-legal bureaucratic practice.[11] These legal domiciles were clearly artificial constructs to some degree. Roman census takers, for example, had to cope with the fact that individuals had property spread in numerous different locations and, by the same token, had dual residences. The legal domicile, Bagnall and Frier observe, was therefore "defined circularly as the place given in the records as one's domicile."[12] Migration and the less permanent wandering of merchants created headaches for officials in both ancient Rome and the Islamic caliphate of Egypt. In both situations, individuals were required to apply to the authorities in order to change their fiscal domicile.[13]

As these sources make clear, the notion of legal domicile as established in ancient Rome or in the Islamic caliphate was a notion tied closely to property or poll taxes, and much the same was true for many parts of northern Europe after the decline of the western Roman empire. I would

---

Citizen," *Speculum* 48 (1973): 694–713; idem, "A Consilium of Rosello dei Roselli on the Meaning of 'Florentinus,' 'de Florentia' and 'de populo," *Bulletin of Medieval Canon Law* n.s., 6 (1976): 87–91; see also Peter Riesenberg, *Citizenship in the Western Tradition: Plato to Rousseau* (Chapel Hill, 1992).

[9] Roger S. Bagnall and Bruce W. Frier, *The Demography of Roman Egypt* (Cambridge, 1994), 14–16, 167–68.

[10] Goitein, *Mediterranean Society*, 2:380–94, where a distinction is drawn between newcomer (tāri') and permanent resident (qātin). Writes an eleventh-century Tunisian merchant and scholar (2:385): "I intend to pass the winter in Jerusalem, for I have learned about the [bad] Nile [which meant famine for Egypt to where the writer expected to travel]. Furthermore, I am registered in the revenue office (*kharāj*) of Old Cairo as resident. Originally they registered me as a newcomer, but when my stay in the country extended, I became a *qātin*."

[11] Immanuel C. Y. Hsü, *The Rise of Modern China* (New York, 1970), 73–75; Kung-Chuan Hsiao, *Rural China: Imperial Control in the Nineteenth Century* (Seattle, 1960), 43–83; E. Herbert Norman, *Origins of the Modern Japanese State*, ed. John W. Dower (New York, 1975), 328. My thanks to Benedict Anderson for bringing my attention to this subject.

[12] Bagnall and Frier, *Demography*, 15.

[13] On restrictions on mobility, see Goitein, *Mediterranean Society*, 4:25 and 4:39. The vizier in 1124 gave an order "to the heads of the police in Cairo and Fustat to register all inhabitants of the capital, street by street and quarter by quarter."

argue that a person's address, at least a person's street address, was not a major element of the identity templates of the record-keeping bureaucracies of antiquity in the circum-Mediterranean region, chiefly because there do not seem to have *been* any identity templates. As Max Weber and his interpreters would argue, this is because these were not rational-legal bureaucracies.[14]

The Egyptian census itself did not commonly use any precise address as an identifying label, and the extant tax records from elsewhere in the Roman empire rarely indicate status, let alone domicile, preferring to use lineage as an identity label.[15] The decentralized nature of Roman administration suggests that local officials simply knew all the individuals within their area of competence. These were patrimonial bureaucracies, and officials did not bother much about establishing the identities of their populations for use by a distant, impersonal bureaucracy. Ancient Roman administrators did have a version of a cadastral map, but used it to attach property to people for purposes of taxation, not the other way around.[16] It is the systematic relationship between identity and address for purposes other than purely fiscal ones that is unique to developing bureaucratic practice in early modern Europe, Ch'ing China, Tokugawa Japan, and surely elsewhere too.

The ubiquitous use of address in modern bureaucratic regimes leads us to the argument that use of the address, and the associated mental habit of attaching identity to residence, is a condition of modernity. The very idea that identity can be attached to geographical location is an essential intellectual component of administrative concerns about bandits, Bedouins, vagabonds, beggars, and other mobile populations, concerns that typically develop in centralizing political jurisdictions.[17]

---

[14] Weber's classic description of rational-legal bureaucracies can be found in Robert K. Merton et al., eds., *Reader in Bureaucracy* (New York, 1952), 18–27.

[15] See Bagnall and Frier, *Demography*; A. H. M. Jones, *The Roman Economy: Studies in Ancient Economic and Administrative History*, ed. P. A. Brunt (Totowa, N.J., 1974), 228–56.

[16] O. A. W. Dilke, *Greek and Roman Maps* (London, 1985), 108–10, 188–89.

[17] Useful discussions of the literature for the early modern period can be found in Karen Barkey, *Bandits and Bureaucrats: The Ottoman Route to State Centralization* (Ithaca, 1994), 12–17 and passim; see also James C. Scott, *Seeing Like a State: How Certain Schemes to Improve the Human Condition Have Failed* (New Haven, 1998), 1–2; Marc Raeff, *The Well-Ordered Police State: Social and Institutional Change through Law in the Germanies* (New Haven, 1983), 88–92; Thomas McStay Adams, *Bureaucrats and Beggars: French Social Policy in the Age of the Enlightenment* (Oxford, 1990), 49. The 1572 Vagabonds Act in England (13 Elizabeth I, c. 5), which enjoined justices of the peace to make a register of all local poor, reflects this localizing ideology; see Norman L. Jones, "William Cecil and the Making of Economic Policy in the 1560s and early 1570s," in *Political Thought and the Tudor Commonwealth: Deep Structure, Discourse and Disguise*, ed. Paul A. Fideler and T. F. Mayer (London, 1992), 171–72.

As this suggests, address-keeping is a habit of nation-states interested in documenting their citizen populations. This is not merely a theoretically inspired perception that we have imposed on the past. Ernest N. Williams records the diary entry of a Swiss traveler named Ludwig Meyer von Knonau who, writing in 1789, could not be plainer about the relationship between the address and the state. "As we got to the other side of Bözberg near Nornussen and entered Austrian territory we were struck by the sight of the house-numbers, which seemed like a kind of shower, and appeared to us as a symbol of the hand of the sovereign, inexorably extending over the property of the private person." As Williams remarks,

> These little white plaques with their street-names and house-numbers, which the Emperor Joseph II imposed on his subjects, also indicate that the Swiss traveller had entered the historical epoch in which we live today: the era of the nation-state. This phenomenon was one of the chief products of the Ancien Régime, comprising the apparatus of bureaucracy on the one hand and the citizen-masses on the other, all equal in their subjection to it.[18]

House-numbers and formal addresses are related, in turn, to cadastral mapping, because it takes a cadastral mentality—the idea that house sites can be clearly identified and their boundaries ascertained by means of the technology of the survey—to make a precise address thinkable.[19] This explains why the numbering of houses in French cities and towns took place more or less contemporaneously with the national map surveys of the eighteenth and early nineteenth centuries. As David Garrioch notes, numbers were first mandated by royal government in 1768. Local governments began even earlier. The municipal archives of Lille contain an ordinance from a few years earlier, December of 1765, stating that the king is requiring the city to number houses in order to facilitate the lodging of troops. The municipal magistrates of Lille encouraged the principle of enumeration, noting that house numbers would help the city to collect the tax known as the vingtième.[20] The royal ordinance itself was

---

[18] Ernest N. Williams, *The Ancien Régime in Europe: Government and Society in the Major States, 1648–1789* (New York, 1970), 1.

[19] Roger J. P. Kain and Elizabeth Baigent, *The Cadastral Map in the Service of the State* (Chicago, 1992), 344.

[20] Archives municipales de Lille, fonds Lillois 22.243, municipal ordinance of Dec. 6, 1765. My thanks to Gail Bossenga who drew this ordinance to my attention.

recapitulated in 1791 and 1805.[21] The two national map surveys of the Cassinis, in turn, were completed in 1744 and 1788, and the Napoleonic cadaster was first organized and planned between 1802 and 1807.[22]

As we have seen, late medieval Marseille did possess what amounts to a set of linguistic cadasters, voluminous records regarding property rights and obligations that were informed by a workable linguistic map and characterized by plat descriptions that, to contemporary users, were readily identifiable. By the same token, Marseille also had addresses, that is to say, descriptions of places used to help establish the identity of an individual, either voluntarily or coercively. We have already seen quite a few in the case of the register of the confraternity of St. Jacques de Gallicia, and here we will turn to identity clauses used in notarial acts, seigneurial rent registers, and court records. They were not always used, and, in general, addresses and other identity labels were not standard requirements of identity clauses in late medieval legal acts: there was no identity template like the one found in 1907 that demanded the use of an address. Yet the custom of taking identities and using identity labels in Marseille, if not fully routinized, was nonetheless present in certain records, and addresses were already being used in ways not wholly anticipated by Roman and Islamic record-keeping bureaucracies.

In the first section I will survey the elements commonly used in all types of fourteenth-century records to create identity clauses—name, profession or status, parentage, and address. The next two sections explore the use of address more specifically in notarial and seigneurial records from the same century. The most important thing to emerge from even so chronologically limited a study as this is that notaries, however much they were instrumental in remapping the space and identity of property sites, were of negligible importance in the process whereby addresses became attached to identities. Instead, the impulse to address came in part from property lords or seigneurial officials. To a small but noticeable degree, this impulse was also a preexisting component of vernacular identity constructs, particularly in the case of artisans, service trades, professionals, laborers, and fisherman. What we have in the case of addresses, then, is another example of a rationalizing process in the nature of identity constructs that took place outside the realm of state interest. This creates a

---

[21] On the ordinances of 1768, 1791, and 1805, see David Garrioch, "House Names, Shop Signs and Social Organization in Western European Cities, 1500–1900," *Urban History* 21 (1994): 37–38.

[22] On the Cassini national map surveys, see Josef Konvitz, *Cartography in France, 1660–1848: Science, Engineering, and Statecraft* (Chicago, 1987), 8–31. On the Napoleonic cadaster, see Konvitz, *Cartography*, 53–62; Kain and Baigent, *Cadastral Map*, 228–31.

problem for the easy assumption that the invention of the address goes hand-in-hand with the development of nation-states. In southern European polities, especially in a city like fourteenth-century Marseille where the state was at best a weak concept, addresses were developed by linguistic communities that were as much private as they were public, especially private and ecclesiastical landlords. This is not an argument that modern states and modern record-keeping bureaucracies do not come to have a vested interest in identity and address. They do. It is only to point out that the roots of the phenomenon cannot be traced to state assertiveness and inventiveness.

SHAPING IDENTITIES

Almost all identifying labels used in notarial acts, court cases, rent registers, and fiscal documents from medieval Marseille fall into six rough categories: name, parentage and/or marital status, legal status, trade or profession, place of origin, and address. The forename was the most basic form of identity in late medieval Marseille, as we can tell from the simple fact that the few extant contemporary indexes to the names of people found in notarial, fiscal, and seigneurial documents were invariably organized alphabetically by forename.[23] In addition to a forename, almost all male citizens of fourteenth-century Marseille used a surname that was passed on from father to children. This was not unusual, for the surname made its appearance all over Europe between the tenth and the fifteenth centuries, and most were fixed by the end of the period.[24] The appearance of the surname is a phenomenon of considerable importance, because although surnames did not necessarily develop to promote ease of bureaucratic identification, they certainly came to serve that purpose.[25] In fourteenth-century Marseille not all surnames were standardized as of yet. A number of men had aliases, some of them charming, as in Peire Borel, alias the Rich (*lo Ric*), perhaps used to distinguish him from Peire Borel, alias the Rascal (*lo Rasquas*), others more pedestrian, as in Rostahn Berart, alias de Mayronas.[26] These aliases were used fitfully by notaries and

---

[23] See, inter alia, ADBR 2HD E7 (1349–1353), 351E 408 (1445–1449), 351E 378 (1450).

[24] The literature is ably surveyed in Roberto S. Lopez, "Concerning Surnames and Places of Origin," *Medievalia et Humanistica* o.s., 8 (1954): 6–16.

[25] For France, see A. Dauzat, *Dictionnaire étymologique des noms de familles et prénoms de France* (Paris, 1951). Françoise Zonabend surveys some of the recent literature on naming and discusses naming as a strategy of classification in her "Le nom de personne," *L'homme: revue française d'anthropologie* 20 (1980): 7–23.

[26] The two Peire Borels are found in ADBR 5G 114, fols. 44v–45r, 204v. "Rostahn Berart,

scribes. Sometimes, however, they mattered a great deal. In one case, a young Jew who converted to Christianity and took the name Peire de Anti-bolo provided his prior name or alias, Astrugon de Vidas, to the notary. Peire/Astrugon was acquitting his father's brother of a debt owed to him, namely, the estate of his own late father.[27]

In Marseille, as elsewhere in France at a comparable time, women's sur-names were somewhat more variable than those of men.[28] Wives were usually known by the husbands' surnames. To give an example drawn from the register of the confraternity of St. Jacques de Gallicia, "lady Calviera Duzes" was identified as "the wife of Sir Bertran Duzes" (*molher den Bertran Duzes*).[29] In this register women often lost their last names entirely and became simply a forename attached to a husband, as in the case of "Peire Bertomieu, in the Fruitery, and his wife Gillaumeta" (*Peire Bertomieu, en la Frucharia, essa molher Gillaumeta*), or "Selita, wife of Pons Folquier."[30] The spelling of the surname was often feminized, and this sometimes involved changes in consonants as well, as in the case of the change from Esteve to Estephana or Albaric to Albariga. It was not uncom-mon for widows to take back their maiden names. In a court case from 1362 pitting Bertomieua Acharda against her daughter-in-law, Mandina Folquessa, both women used their maiden names because their husbands, Raymon Bermon senior and Raymon Bermon junior, were dead.[31] A few wives used both, as we can tell from situations where a woman gave her husband's surname and added an "alias" followed by her father's surname, sometimes reversing which name was the "alias" in another act. In one act, Biatris Bertrana described herself as the daughter of Peire Bertran and Jacma Coffarda and indicated that she herself was also known as Biatris Coffarda; from an earlier act we learn that her uterine half-sister Pellegrina also bore their mother's surname, Coffarda.[32] This and similar examples show that although use of the father's surname was the norm, it was not a hard and fast rule.

The practice whereby a child took his or her father's forename as a

---

alias de Mayronas" can be found in several records; see, for example, the casebook of the notary Peire Giraut, ADBR 381E 81, fols. 22v–23r, 8 June 1358.

[27] ADBR 355E 3, fol. 105r, 11 Oct. 1350.

[28] See the discussion of female anthroponymy in Monique Bourin and Pascal Chareille, eds., *Désignation et anthroponymie des femmes. Méthodes statistiques pour l'anthroponymie*, part 2 of *Persistances du nom unique*, vol. 2 of *Genèse médiévale de l'anthroponymie moderne* (Tours, 1992).

[29] ADBR 2HD E7, p. 4.

[30] Ibid., p. 8, 10.

[31] ADBR 3B 64, fols. 65r–71r, 10 Nov. 1362.

[32] ADBR 381E 44, fols. 131r–v and 132v–133r, 22 Nov. 1347.

surname is rarely found in Marseille, although it was common enough elsewhere in Mediterranean Europe. Most surnames that are obviously patronymics in origin (e.g., Johan or Johannis as a surname) had, by the mid-fourteenth century, become fixed surnames, passed from father to child. This process, of course, mirrored the petrification of the cartographic landscape as discussed in previous chapters and the increasing separation between signifier and signified.

About as uncommon as surnames taken from the father's name were surnames or identifying labels taken from the craft or trade currently being practiced by the person in question. In most cases, again, such names had become fixed surnames, and in most cases the person in question was no longer practicing the craft. Consider the men bearing the name *Faber* or *Fabri* (Smith) in mid-fourteenth-century Marseille. In addition to one smith, one fastener, and two goldsmiths, all of whom are arguably smiths of one sort or another, there was an apothecary, a squire, a schoolmaster, a vintner, a jurist, a notary, a shepherd, a Carmelite brother, a deacon, two saddlers, two clerks, two bakers, two merchants, two monks, two shoemakers, three butchers, three gardeners, four priests, four fishermen, five seamen, and twelve laborers. Fixed surnames based on craft were declined in the genitive in Latin (*Piscatoris, Fusterii, Fabri*). A surname not in the genitive, such as *Petrus Curaterius*, is best translated as "Peire the Cobbler." It almost certainly means that Peire was a cobbler. There were no declensions in the spoken Provençal of the city, so in Provençal, Peire would have been known as "Peire Curatier." We must assume, in cases where the notary was pointedly using the nominative form, that he had heard something different from the person being identified, perhaps a Provençal article not used in Latin, as in "Peire lo Curatier."

One example of both patronymic and craft-based usage is that of Aymar Tibaut (*Aymericus Tibaudi* in Latin). Aymar's father was Tibaut Pastissier or Tibaut the Pastry Chef (*Tibaudus Pastisserius*). His son Aymar had taken a patronymic as his surname. Aymar, furthermore, was himself a pastry chef like his father.

In a very few cases, people seem to have exchanged their surname for a trade designation simply to avoid confusion. One person, named Peire Bermon, sometimes called himself "Peire the Glassmaker" to distinguish himself, perhaps, from other Peire Bermons, such as the Peire Bermon who sometimes called himself "Peire the *Mosclalherius*." Both had children who always bore the surname Bermon (*Bermundi*).

Toponymics (e.g. *de Aquis, de Tholono*) are harder to trace, since one cannot usually tell if the person in question was from that town or merely bore the surname of an ancestor.[33] Immigrants frequently shed surnames and took on toponymics that were then passed on to their children. Yet another Peire Bermon (there were at least thirteen Peire Bermons in the city), a knight from the town of Sant Felis, passed on a new surname "de Sant Felis" to his children Anselm, Jacme, and Ricava.[34]

The use of a fixed surname of any type by people of common status dates to at least the mid-thirteenth century in Marseille. It probably arose in the context of a developing notarial or scribal culture that found itself in need of more precise identifying labels.[35] In this consistency of surname usage, Marseille—and for that matter other regions of southern France, Mediterranean Europe, and England—stood in sharp contrast to Florence where 63 percent of the population at the time of the Catasto of 1427–30 still used toponymics or patronymics that changed from father to son.[36] Herlihy and Klapisch suggest that "the appearance of a collective, permanent cognomen used to designate a group of kin is tightly associated, in Tuscany, with the appearance of the lineage."[37] The use of the fixed surname, in Marseille and elsewhere in southern France, clearly had different origins, related less, I think, to the appearance of the lineage and more to the spread of a particular type of notarial culture. We cannot assume that naming patterns reflect social structures in a straightforward and uncomplicated way without taking into account the influence of the culture of record of a given area.

Use of the surname, then, was nearly universal, and since there were

---

[33] See Richard W. Emery, "The Use of the Surname in the Study of Medieval Economic History," *Medievalia et Humanistica* o.s., 7 (1952): 43–50.

[34] The thirteen included a canvas-maker, a butcher, two laborers, two *mosclaherii* (father and son), two glassmakers (father and son), two merchants, a knight, a notary, and a priest. The knight was an important landlord; although he died in 1337, his name, and those of his children, continued to show up in numerous sales and other property transactions up to 1356. He was always called "Petrus Bermundi, de Sancto Felicio," even in documents where his children were called "de Sancto Felicio." See, for example, ADBR 391E 11, fols. 139v–140r, 3 November 1337; 358E 86, fols. 94v–95v, 7 August 1354; and 355E 292, fols. 57r–58v, 29 August 1356. In other documents too numerous to cite, his children always took the name "de Sancto Felicio."

[35] This is to judge by the types of surnames found in the register of the notary Giraud Amalric; these include a mixture of nominative and genitive case names. See John H. Pryor, *Business Contracts of Medieval Provence: Selected "Notulae" from the Cartulary of Giraud Amalric of Marseilles, 1248* (Toronto, 1981), and Louis Blancard, *Documents inédits sur le commerce de Marseille au moyen âge*, 2 vols. (Marseille, 1978 [1884]).

[36] David Herlihy and Christiane Klapisch-Zuber, *Les toscans et leurs familles: une étude du catasto florentin de 1427* (Paris, 1978): 539. English surnames were generally hereditary by 1300, a process that C. M. Matthews assigns to the effects of the king's new legislation and wide taxation; see his *English Surnames* (New York, 1966), 48–58. For southern Europe, see Lopez, "Concerning Surnames."

[37] Herlihy and Klapisch-Zuber, *Toscans et leurs familles*, 537.

more surnames than forenames, they did much of the work necessary for a reasonably precise identification. Other labels were used less consistently. Following the name, one almost never sees personal characteristics (i.e., "the fat one") that were not fixed surnames, although from time to time one does find people described by their nationality (the Catalan, the Frenchman). "Senior" and "junior" were usually used in cases where father and son bore the same name. Age was also conveyed by diminutives, typically the suffix "et," for both men and women; the habit is found among both Christians and Jews and across the social spectrum. Men typically dropped their diminuitives in their late teens or twenties—a Johannetus or Johanet would become Johannes or Johan, depending on the language of the record, and Monnetus or Monet would become Raymundus or Raymon. Women often kept their diminuitive names longer, so that one can find an Esmenjardeta or a Biatriseta who is clearly an older woman. Unemancipated minors under the age of twenty-five who appear as clients in notarial records usually had to give their approximate age ("between twelve and eighteen years of age"), and from this information it would be possible to discover whether there was, in Marseille, an age at which names customarily lost their diminuitive forms.

The patronymic in Marseille had, in a sense, evolved into the expression *filius* or *filia*, followed by a name in the genitive, meaning "son of" or "daughter of." Women were commonly identified in records of all types as a daughter or as a wife or both, probably since most were not considered practitioners of a trade; hence, a woman might be known as "Esmenjardet Duranta, daughter of Giraut Faber and wife of Isnart Durant" (*Esmenjardeta Duranta, filia Giraudi Fabri et uxor Isnardi Duranti*). The ancestor was usually a father, although a notable woman could from time to time be the favored ancestor. More distant relationships, such as niece or cousin or grandfather of someone can also be found in notarial acts, but these were used only if pertinent to the act, and do not establish identity so much as they describe the capacity in which the individual is acting ("as uncle and legal administrator of so-and-so"; "as heiress to her uncle so-and-so").

Name and relation were related to the familial identity of the person. Status, trade, or profession indicated something else, such as the social role or roles into which the person fit. Status, as I have used the word here, is an umbrella term designating categories of people who had a legal status that differed in some way from the norm: Jews, slaves, members of the clergy, and nobles. Jews could not serve as witnesses to notarial contracts and had other legal disabilities. Clerical status was a distinct legal category, since clerics could not be tried in secular criminal courts. In Marseille, nobles did not have any distinctive legal privileges, apart from the privilege of bearing noble titles, although many of them apparently thought they did, refusing, for example, to pay taxes. There was no particular name for someone who belonged to the non-noble free Christian population and, as in other Mediterranean cities, few if any formal distinctions between enfranchised burghers and unenfranchised citizens. Everyone, including Jews (but excepting slaves), could be a citizen of Marseille once certain basic requirements were met. Ordinary free citizens, such as laborers, fishermen, artisans, merchants, retailers, physicians, and notaries frequently identified themselves by means of their trade or profession, and used a trade or profession in much the same way that a nobleman would use "knight" (*miles*) or a Jewish woman would use "Jewess" (*judea*). This seems to suggest a certain equivalence between these categories. Yet it is worthwhile keeping status distinct from trade or profession, if for no other reason than that trade or profession did not have any legal quality. In addition, one could be a noble *and* practice a trade such as merchanting or banking. A number of Marseille's Jews worked in trades of various sorts, a situation that resulted in identity clauses such as "Duranton de Castelnau, Jew, dyer, and citizen of the lower city" (*Durantonus de Castronovo, judeus, tenchurerius et civis ville inferioris*).[38]

Identity clauses sometimes included place of origin, an important legal category which was used by people who were not citizens or residents of the city, such as Marseille's mobile population of foreign merchants. It was also used by recent arrivals who had not yet become citizens of the city and still thought of themselves as tied to another locale. Citizens and residents of Marseille often, but not always, mentioned this fact—the common expression was *civis et habitator Massilie*—because to be a citizen was to have important legal rights and responsibilities. Often a notary made a distinction between the citizen-resident and residents who were not citizens by giving a person's legal place of citizenship or residence ("so-and-so of the castrum of Berre, now a resident of Marseille").

Addresses, when used, were given in forms hardly distinguishable from the notarial site clauses discussed in chapter two, although they never included abutments. The only important linguistic difference between addresses and site clauses is that the latter use the word *sitam* or its equivalent, as in *quadam domum sitam in carreria Negrelli*, whereas addresses use the words *morans* or *degentem*, as in *Anthonius de Aquis, morans in carreria*

[38] ADBR 381E 73, fols. 50r–v, 12 Aug. 1342.

Negrelli ("Antoni d'Ays, residing in Negrel street"). Addresses invariably referred to the lowest level of cartographic awareness, namely streets, vicinities, or landmarks. Even more than property sites, addresses were never nested within any of the city's administrative districts, as in a hypothetical "Antoni d'Ays, living in Negrel street and in the sixain of Drapery." A number of identity clauses do mention the upper or lower city, but these are typically introduced not by *morans* or *degentem* but instead by the preposition *de*, as in *Anthonius de Aquis, civis et habitator de ville inferioris Massilie* ("Antoni d'Ays, citizen and resident of the lower city of Marseille"). This means that the phrase was intended to denote a legal status—citizenship in the lower city—and not an address.

The identifying label given to individuals in Marseille's records was some permutation of all, some, or none of these individual identity labels. Trade or profession and status were the most common labels, but patterns changed considerably from one type of record to another and, in the case of notarial casebooks, from one type of act to another. What is most important is that, as we have seen elsewhere, although each type of record and act had characteristic tendencies, in no case was there a formula or template that structured identities and led to consistent patterns in identity clauses across the record. Take the example of the register of the confraternity of St. Jacques de Gallicia, a few entries from which follow:

> Galtelme Malet maelier (Gautelme Malet, butcher)
> Guilhem Boquier mazelier (Guilhem Boquier, butcher)
> Dona Alaeta Toarda en la Frucharia (Madam Alaeta Toarda, in the Fruitery)
> Johan Rafel en la Sabataria de Sant Jacme (Johan Rafel, in the Shoemakery of St. Jacques)
> Dona Gillaumeta molher de Peire Bertomieu en la Frucharia (Madam Gillaumeta, wife of Peire Bertomieu, in the Fruitery)
> Matieu de Vals
> Quatarina Rebolla
> Dona Belisen pastissiera que esta en la Frenaria (Madam Belisen, pastry chef, who is in the Jewelery)[39]

A profession, that of butcher, was given by the first two men in this example, Gautelme Malet and Guilhem Boquier, although the profession itself was spelled two different ways by the scribe. Alaeta Toarda, who styled herself as a "Madam," came next, and she gave her identity by means of an address in the Fruitery, as did the man who followed, Johan

[39] ADRB 2HD E7, p. 27.

Rafel—*en la*, here, is the Provençal equivalent of the notarial *morans in*. Madam Gillaumeta, the next entry, also lived in the Fruitery, but she saw fit to identify herself also as the wife of Peire Bertomieu and used this in place of a surname. The man and woman who followed, Matieu de Vals and Quatarina Rebolla, gave themselves no identity labels at all; Quatarina, here and elsewhere in the register, apparently did not think of herself as lady-like enough—or was not considered by the scribe to be lady-like enough—to be a *dona*. They were followed by Madam Belisen, who lived in the Jewelery, a center of the pastry trade. We cannot tell from this record whether Belisen was a *pastissiera* (pastry chef), or whether she considered this her surname, or both.

The absence of a universalizing identity template here and in other types of records from mid-fourteenth-century Marseille is typical of a record-keeping culture that is not centralized. Confraternal scribes and public notaries may possibly have been aware of the problem of ambiguity and may have asked their clients to provide identity labels. But even if they did, they did not consistently record the responses.[40] The reason may be that in many cases, even in a large city like Marseille, scribes, notaries, and the officials who used their services were acquainted with the people with whom they dealt. In addition, their scribal tradition was rooted in the past, in an age when the city was smaller, when it may have been possible for a notary's or an official's range of acquaintance to encompass a good percentage of the legally active individuals in the city. Within such a community of knowledge, identity labels served as mnemonic devices for notaries, not as formal, impersonal markers of identity. The mnemonic chosen depended heavily on notarial expectations and mnemotechnics, although the very diversity of identity clauses suggests that clients had some latitude to identify themselves as they so chose. Since the science of memory was so well developed in the middle ages, there is every reason to believe that notaries were at least to some extent conscious of the legal role of memory in the act of identification.[41]

Some of these identity clauses included addresses. In all types of records from mid-fourteenth-century Marseille, addresses were used more commonly in clauses identifying laborers, fishermen, artisans, and members of professions, and much less often in clauses identifying nobles, Jews,

---

[40] *Pace* Emery, who suggested some years back that a notary "took elaborate pains to avoid confusion"; see "Use of the Surname," 48.

[41] Mario Montorzi, *Fides in rem publicam: ambiguità e techniche del diritto comune* (Naples, 1984), 215–66; Mary Carruthers, *The Book of Memory: The Study of Memory in Medieval Culture* (Cambridge, 1990), 8.

and merchants of great standing. In some cases, as we shall see in the following two sections, it is clear that ordinary men and women were simply more likely to link identity and address. In most cases, however, addresses were demanded by concerned creditors or powerful landlords. In either case, addresses were being attached to identities in interesting ways by record-keeping bureaucracies working independently of state interests.

IDENTITY AND ADDRESS IN NOTARIAL CASEBOOKS

The identification practices of Marseille's public notariate do not show a strong association between identity and address. Insofar as we can distinguish notarial preferences from those of their clients—a tricky matter—it seems as if notaries generally preferred status or profession as identity labels. Beyond this, certain types of notarial acts demanded more precision in identity clauses than did others. But although notaries may have had a preference, sometimes a preference imposed by the nature of the contract, they did not, as argued above, routinely apply any template to their clients' identities. One result of this is considerable variation in notarial identity clauses, a variation that probably reflects the different vernacular identity constructs characteristic of given social groups. Put differently, vernacular identity constructs pushed through notarial preferences to some degree. As far as this book is concerned, the most significant feature of vernacular identity constructs is that members of the non-noble free Christian population, such as artisans, retailers, and laborers, were more prone than any other group to identify themselves by means of addresses.

To begin, patterns of identification in notarial acts in mid-fourteenth-century Marseille can vary noticeably with the type of act and the type of clause within the act, a variation explained in part as a function of the legal requirements of the act. Take the case of parentage. In the several hundred dotal acts extant from the mid-fourteenth century, the groom's father was named 60 percent of the time, and the bride's father was always named, even if, in most cases, the father was already dead. This is because the dowries and inheritance practices were highly regulated by statutes, and parentage was an exceedingly important legal component in this kind of act. In other acts, the percentages are different. Quittances are a particularly good category of act to use for purposes of comparison because they mark the end of a relationship, not the beginning, and for this reason their future significance is much less when compared to loans, dotal acts, testaments, or business societies. Around 700 quittances have survived from the mid-fourteenth century; in these acts, use of parentage is much less frequent than it is in dowries, varying according to the status, gender, and age of the clients but averaging around 16 percent. Many identity clauses referring to parents, moreover, did so because they were identifying young children.

The structural relationship between the contracting parties in notarial acts also shaped the nature of identity labels; here again, it looks as if notaries (or interested clients) imposed their own standards on identity clauses. In noncommercial loans, for example, borrowers were more carefully identified than were lenders. In one noncommercial loan involving the relatively large sum of thirty pounds, the debtors were identified as Jacme and Duranta de Sanoya, he a fisherman, both residing in the street of the Oarmakers, along with another fisherman, Pons Alvernhacii, and his wife Simona. The creditor was thinly identified as Boniuse Mosse, a Jew. In another loan, Guilhem Gabian, carefully identified as a laborer and a resident of the suburb of St. Catherine, borrowed seven pounds and ten shillings from a woman identified simply as Raymona Fornilheria.[42]

Last, all clients in notarial acts were more carefully identified than the witnesses listed at the end of the act. Using quittances again, 71 percent of the 1586 clients were identified by more than just a name, whereas only 48 percent of 1310 witnesses were identified by more than name. Witnesses to notarial acts were almost never brought before judges or arbitrators to testify to what they had seen, and since their identities did not make much practical difference, less ink was wasted on their identities. To conclude, in drawing up a legal act, the notary routinely elicited identification labels from all participants in the act. Witnesses may have been asked to give nothing more than a forename and surname. In contrast, clients were more often asked to provide additional labels, especially in cases where such labels might serve an important legal purpose.

The legal requirements of specific acts most often affected identity clauses when kinfolk were involved. In such acts, the name of the individual might be followed by "father of," "uncle of," "nephew and legitimate heir of," and so on. In these cases, we are dealing not with a routine or generic identity. Instead, the notary was being careful to identify what facet of a person's legal persona was implicated in the act. In a quittance

---

[42] See, respectively, ADBR 391E 11, fol. 25v, 11 May 1337; ADBR 381E 384, fol. 204v, 24 Oct. 1337.

from 1350 in which Jacmeta de Batut's name is followed by "wife of the late Marques de Batut and mother of Bernat de Batut," Jacmeta may well have thought that her status as a widow of Marques was an important aspect of her generic identity, but surely she did not think being a mother of Bernat had anything to do with it.[43] In other acts from the period— she appears in ten—Jacmeta mentioned Bernat or her other children only in those acts involving their interests; otherwise she was satisfied with "daughter of" (a single instance in an early act), or "wife of," or "lady" (*domina*), or no label at all. "Mother" was mentioned in the quittance from 1350 because she was acquitting her son, Bernat, for a debt of thirty-six and a half florins, almost certainly part of the dowry that had been paid over to her late husband.

It is difficult to know where to draw a line between a generic identity and legal persona, and possibly anachronistic to assume that contemporaries drew such a line. Yet the issue is important. One of the effects of the *ius commune* was to fragment the individual into a multitude of legal personae. To give one example, a person could own or manage an estate in a variety of ways—as a father for an unemancipated son, as a husband for a wife or as father-in-law for a daughter-in-law, as a guardian for a ward, as an heir for life, as a member of a partnership, as a lord, or as a pro-prietor. The list of possibilities goes on. In any contract involving the estate, a person acted only in the given legal capacity. Notaries routinely defined this capacity by means of appropriate labels. Did this fragmentation of legal personae and the multiplication of identity labels have any effect on the ways in which people construed their own identities? This is a question of significance in the context of the identity industry that has sprung up in the last decade or so, and it seems to me that there is a great deal of fruitful work to be done with the legal categories developed by the *ius commune* and brought to consumers of the law by notarial practice.

Although the nature of their acts sometimes encouraged notaries to provide identity labels, this does not necessarily mean that notarial identity templates invariably pushed aside or overrode vernacular identity templates. Nor does it mean that a carefully constructed identity was in any way a legal norm. Consider that perhaps the most surprising feature of notarial records from mid-fourteenth-century Marseille is the degree to which people were not identified at all. For someone like the great merchant Peire Austria, the only man (other than his father) to bear that name in mid-fourteenth-century Marseille, it is perhaps understandable

[43] ADBR 355E 2, fols. 153v–154r, 20 Jan. 1350.

that he, or the notaries who drew up acts in which he was involved, did not bother listing his trade, parentage, or address too often.[44] In other cases it was not so simple. According to the prosopographical index, the city was home to several dozen Peire Martins in any given decade in the mid-fourteenth century. We can confidently identify a jailer, cleric, draper, baker, mariner, merchant, notary, gardener, fisherman, priest, shoemaker, nobleman, squire, squire and banker, weaver, a servant in the episcopal house, and possibly as many as eleven different laborers. Some of these Peire Martins lived on the street of the Brassery, the street of the Jerusalem, the street of the Dominicans, the street of St. Jacques, the Fruitery, the Spicery, the street of the Upper Grain Market, the Frache, the street of the Steps, the Goldsmithery, the Tripery, the street of Johan Pedagier, and in the island of Guilhem Sard, near the episcopal house. Among them were the sons of Bertomieu, Gilles, Uguo, Jacme, Laurens, Felip, Raymon, Pons, two Peires, and three different Guilhems. At least six were recent immigrants. The possibilities for ambiguity were rife. Yet this did not necessarily encourage any of these Peire Martins to identify themselves with great precision to the notary, nor did it encourage officious or conscientious notaries, faced with name duplication, to ask for more labels. A draper and self-styled *nobilis* named Peire Martin, for example, practiced moneylending on the side, and as a result crops up frequently in records from the 1350s. He did a great deal of business with the notary Johan Silvester and appears dozens of times in the notary's four extant casebooks.[45] In these casebooks Peire was frequently identified as the son of the late Bertomieu, and once or twice the label "draper" and "residing in the Goldsmithery" was also added to his name, but from time to time the notary simply identified the man as *Petrus Martini*.

Why the absence of any label in a case of such potential ambiguity? The answer to this is simple: notaries in Marseille just knew their clients. This, at any rate, is what we can gather from the fact that people well known to a notary were not identified as carefully as were those who were not public figures or who fell outside the notary's range of acquaintance. Similarly, a person appearing in a series of acts in the same casebook was less likely to be identified as fully as the series progressed, rather like the academic style of citing a text in footnotes. Peire Martin, the moneylender, acquired an identity simply by being a client of Johan Silvester. His iden-

---

[44] Numerous references to Peire Austria can be found in the registers of the notary Peire Giraut (e.g. ADBR 381E 77–83).
[45] See ADBR 358E 84–87, which cover some of the years between 1351 and 1362.

tity was built up in the context of Johan's practice, not on the basis of any individual acts. Thus, the notary himself, as the node of a huge social network, was a powerful element in establishing identity. It was his own character, trustworthiness, and prodigious memory, aided by his written notes, that gave authenticity and legal force to the identity clauses he drew up. This being so, notaries used identifying labels more as mnemonic devices than as elements of a rational-legal science of identification. The mnemonic did not necessarily matter all that much; notaries preferred status or profession, but were open to other possibilities.

The fact that members of specific social groups were typically identified in characteristic ways is the most important indication that vernacular identity constructs could, at times, push through notarial preferences and legal requirements. Consider individuals who styled themselves as noble. In notarial acts, these nobles often mentioned their title and often gave their fathers' names. In quittances, for example, out of 141 clients readily identifiable as nobles, 43 or 30 percent named fathers or, in a few cases, mothers. Women often mentioned their husband's title along with their own lineage. Very few readily identifiable nobles did not use either their status or their lineage, although it would take painstaking prosopographical analysis to identify everyone who had the right to noble status but did not use it. Several examples illustrate the noble pattern. In a loan drawn up by the notary Paul Giraut in August of 1337, Guilhem de Montoliu called himself a squire (*domicellus*) and the son of the late Folco, also a squire. In another loan, from May of 1347, Johan Vivaut described himself as a squire, the son of the late nobleman Berengier Vivaut. Naturally, nobles could use one label or the other. In May of 1337 Carle Athos, who could legitimately call himself a squire, described himself merely as the son of the late Alfans; a month earlier, a relative, Johan Athos, called himself a knight (*miles*), making no mention of his father.

One can see in the mention of the title a measure of pride, because in most cases, at least in fourteenth-century Marseille, noble status made no legal difference. The use of "son of" took on a certain basic importance among the nobles and the mercantile elite as a way to avoid ambiguity. The elite were more successful than members of the general populace at bringing their sons through the hazardous early years of childhood, and since they often recycled the same set of forenames in successive generations, the possibilities for ambiguity were strong. We can see this in the case of the well-known Jerusalem family, a family of armateurs and nobles. In the mid-fourteenth century there were two members of the family named Guilhem, distinguished most easily by their parentage. One was the son of

Marin, the other the son of Uguo. We find in turn two Uguos, one the son of Peire and the other the son of Vivaut. There were at least three Johans, sons of Peire, Marques, and Marin. The Peires numbered four, sons of Peire, Uguo, Bernat, and Marin. The need for clear identification aside, this pattern of using "son of" also reflected a greater consciousness of ascendants and lineage. The recycling of family names, as Christiane Klapisch-Zuber notes, is proof enough of this consciousness.[46]

Nobles rarely used other labels. A few named their rural estates, and a few who were bankers mentioned that fact. But a particularly conspicuous habit of nobles was their reluctance to make any reference to their place of residence. The only exception was Peire Martin, a banker (not the draper, discussed above) who also, on occasion, was called noble. This Peire Martin lived on the street of the Jerusalem, and the sobriquet *de carreria de Jerusalem* became attached to his name in a regular way. This is perhaps because there were so many other Peire Martins in the city, but may also result from the wonderment at seeing a Martin living on the street of the Jerusalem—members of one branch of the Martin family were close allies with the Vivaut in their bitter feud with the noble Jerusalem lineage. So consistent is the noble refusal to give an address that, if it were not for registers of rents and property conveyances that named property sites and listed the names of adjoining property holders, we would have almost no idea where the nobility of medieval Marseille lived. This is not to say that Marseille's nobles did not have a vested interest in a house or compound associated with their lineages, so typical a feature of the domestic life of Florentine magnates or the Genoese nobility.[47] It is only to say that nobles in Marseille simply did not have a sense of an address as a marker of identity in legal contexts. Those who possessed both urban and rural estates may have been unwilling to associate themselves too tightly with their urban property. For others, status and lineage were clear enough.

Jews were much less inclined than nobles to give parentage (around 8 percent did). Jews were frequently identified by the word "Jew" (*judeus* or *judea*). The label was not always given, however, presumably because Jewish forenames—Aaron, Abram, Alegra, Arfila, Astrug, Bondavin, Bonhora, Comprat, Crescas, Jacob, Mira, Mosse, Samiel, and so on—were so distinctive that no mistake could be made about the identities of those

---

[46] Klapisch-Zuber, "The Name 'Remade': The Transmission of Given Names in Florence in the Fourteenth and Fifteenth Centuries," in *Women, Family*, 283–309.

[47] Some of the literature on this is discussed in Jacques Heers, *La ville au moyen âge en Occident* (Paris, 1990), 224–31, 256–58.

who bore them.[48] Of 176 Jews named in quittances (many of them more than once), 23 percent were not labeled *judeus* or *judea*. Almost all Jews had forenames and surnames; many of the surnames, notably the toponyms, are indistinguishable from Christian names. The occasional Jew gave a profession, such as clothier or haberdasher (*corraterius*), dyer, butcher, or parchment-maker, but, as with members of the nobility, this was uncommon. Jews who lent money apparently did not consider this a nameable profession. Quite a few Jews named a place of origin outside Marseille, and many Jews identified themselves as citizens of Marseille (*civis Massilie*). Most important, Jews in Marseille, like members of the nobility, almost never identified themselves by means of an address; hence, in the hundreds of references to Jews in all the loans and quittances from the period 1337 to 1362, there is only one instance in which a Jew, Boniuse Durant, was identified by an address, the suburb of Syon.[49] In a typical loan, a Christian named Uguo Mira, identified as a fletcher living in the Pilings, joined with his wife Plazentina in acknowledging a debt to Boniuse Durant, identified solely as *judeus*.[50] Jewish debtors were no more carefully identified than were Jewish creditors. Typical is a loan from 1358 in which a Christian named Guilelmeta de Orto, identified as a resident of Tart street, lent ten florins to Rosset Abram, identified as a Jew and a son of Mosse Abram.[51] In another from 1342, Durant de Jerusalem, identified as a squire, lent twenty-four florins to Vital Cordier, identified as a Jew and as a citizen and resident of the lower city of Marseille.[52]

This avoidance of address is not just because all Jews lived in the Jewry— a number, like Boniuse, lived elsewhere in the city. It is not because they did not own property—they did. A number of streets in the Jewry even acquired the names of their most distinguished residents, such as the alley of Bondavin, the street of Astrug Cardier, and the street of Salvi de Cortezono. The Jewish fountain, the street of the Great Synagogue, the street of the Jewry, the island of the Upper Jewry, the portale of the Jewry— all these toponyms derived from Marseille's Jewish community were inscribed on the map of the city and all were accepted by Christian notaries and landlords. But Jewish identity was tightly bound up in culture and to a lesser degree in lineage; address was insignificant. As with the

---

[48] Only two given names in mid-fourteenth-century Marseille, Durant and David, were shared by Jews and Christians.
[49] ADBR 355E 36, fol. 34r, 20 Apr. 1360.
[50] Ibid., fol. 121r, 25 Nov. 1360.
[51] ADBR 355E 9, fol. 64v, 17 Aug. 1358.
[52] ADBR 381E 73, fol. 75r, 20 Dec. 1342.

nobility, were it not for site and abutment clauses in records of property conveyances, we would have no idea of where the Jews in Marseille lived. That both nobles and Jews shared an aversion to the address is a curious state of affairs; as Orest Ranum has pointed out to me, it may well reflect a feeling shared by members of both groups that they aren't really citizens with addressable civic identities.[53]

Members of the clergy had perhaps the simplest identification pattern of all: most listed their clerical grade or office (cleric, deacon, priest, canon, monk, prioress, abbess, mendicant) and, in the case of monks, mentioned their monastery. A few female religious dropped their surnames. Novitiates or female religious might mention their family ties but only where this mattered, as in a quittance acknowledging receipt of a legacy; otherwise, lineage was clearly eschewed. The canons of Marseille's cathedral were the only ones to depart significantly from the mold, since many listed their place of domicile (almost always in the street of Francigena or the street of the Stone Image).

Artisans, laborers, retailers, and members of professional groups followed a slightly different pattern. Having no legal status comparable to "noble," "Jew," or "cleric," members of the non-noble free Christian population of the city used labels based on trade or profession instead. A great many, in quittances, were given no labels at all, 426 out of 1212 total clients, or 35 percent.[54] Among the 786 who did have identity labels of some kind, 360 or 46 percent mentioned a trade or profession. Most single women did not identify themselves with a trade or profession, so usage is higher among single men, 63 percent of whom mentioned a trade or profession. Use of lineage or parentage was less common—186 out of the 786, or 24 percent—and the figure is skewed because the great majority of these were children orphaned by the plague who listed lineage as a matter of legal necessity, not, as among the nobility, as an essential component of identity. Among adults of common status, the figure is closer to 10 percent, compared to 30 percent for the nobility and 8 percent for Jews. As a rule of thumb, lineage was more commonly used by the top and bottom of the social scale. Aspiring merchants and rich drapers, conscious of the association between nobility and lineage, named their fathers very often. So did laborers and recent immigrants, possibly because they did not have a clearly defined trade group with which to associate themselves. Artisans, retailers, and members of professional groups, in contrast, based their identity more consistently on trade or profession.

[53] Personal communication, November 1998.
[54] I have included couples acting jointly in this figure.

As mentioned earlier, one of the most striking features of the identity clauses found in notarial acts, regardless of social grouping, is the relative scarcity of addresses. Addresses occur in only 3 percent of the identity clauses in notarized quittances. Sometimes the address stands alone; sometimes we find it in combination with other labels. Women were just as unlikely to use an address as were men. Addresses are vanishingly rare where witnesses are concerned. Of the 2140 witnesses identified in the acts of Peire Aycart between 1349 and 1362, only two had addresses located within the city of Marseille.[55] I suggested above that notaries used identity labels, when they bothered to use them at all, as mnemonic devices. If this is so, then the scarcity of addresses suggests a crucial feature of notarial culture. Notaries typically did not attach identities to the urban landscape, and may have been disinclined to use addresses even if their clients voluntarily offered them. Notaries saw in their clients such qualities as noble status, Jewishness, descent, and trade or profession. Like their colleagues in the larger metropolitan areas of Florence, Genoa, and Venice at a comparable point in time, Marseille's notaries did not see, in their clients, people attached to fixed domiciles.

What this suggests is that when people *did* give addresses, they did so not because the notary asked them to, but instead because they considered the address to be a part of their generic identity. As a result, it is important to ask who did use addresses in the rare cases where they crop up. The answer is that addresses, where found in notarial casebooks, were almost always used by members of the non-noble free Christian population.This habit is most pronounced not in notarial casebooks but, as we have seen, in the register of the confraternity of St. Jacques de Gallicia, where two-thirds of the members identified themselves, at some point in the register, by means of an address. It is likely that some of the scribes who kept the register, especially those responsible for the entries from the year 1352 and after, insisted for unknown reasons that members of the confraternity provide addresses.[56] Identity clauses were not demanded or copied down nearly so consistently in the year 1349. Of the 192 people identified, eighty-nine were identified by name alone, ninety-one by name and occupation, eleven by name and address, and one by name, occupation, and address. As the variation suggests, this scribe did not insist that confraternity members provide identifying labels, although one hundred and six seem to have provided them as a matter of course. One label was

[55] ADBR 355E 34–36, 290–93.
[56] See ADBR 2HD E7, pp. 69 and following.

Table 5.1. Frequency of address use in notarial quittances and loans, 1337–62

	N	Clients with addresses (%)	Clients without addresses (%)
Quittances	1609	3.1	96.9
Loans, creditors	820	3.5	96.5
Loans, debtors	881	30.5	69.5

Sources: 1609 notarial quittances and 1701 notarial loans in ADBR 300E 6; 351E 2–5, 24, 641–45, 647; 355E 1–12, 34–36, 285, 290–93; 381E 38–44, 59–61, 64bis, 72–87, 393–94; 391E 11–18; AM 1 II 42, 44, 57–61.

generally enough, occupation being preferred to address. Nonetheless, twelve people, 6 percent of the total, did volunteer an address. As this shows, an address was a more significant element of vernacular identity constructs than notarial usage alone would suggest. Since notaries favored status or profession as labels, it is even possible, although unprovable, that in constructing identity clauses notaries routinely discarded the addresses that their clients may have offered voluntarily.

This makes sense. The artisans, retailers, and laborers who constituted the bulk of the membership of this confraternity, as we saw earlier, tended to think of the city as a set of vicinities defined as units of sociability rather than as units of cartography. These units of sociability were easily available as markers of identity, and because these vicinities carried status they could fulfill much the same role that lineage did in the case of the nobility or ethnic identity for the Jewish population of the city.

There is one major exception to this general rule of notarial disinterest in addresses: around 18 percent of the identity clauses in notarized noncommercial loans made between 1337 and 1362 mention an address, six times more often than in quittances. Curiously, usage of other identity labels in quittances, namely trade or profession, is *not* significantly different from the pattern typical of loans: around 59 percent of single men involved loans mentioned trade or profession, compared to 63 percent in the case of quittances. Why the remarkable increase in addresses in notarized loans? The answer is obvious: the vast majority of the addresses used in loans—269 or 90 percent of the total of 298 addresses—are found in clauses identifying non-noble Christian debtors, and 30.5 percent of the debtors in loans are identified by means of an address (Table 5.1). A quittance marks the end of a debt; in contrast, a loan marks the beginning; hence, it was the creditor in these cases who in all likelihood asked the notary to include the debtor's address the more easily to track him or her down in case of default. Here, the creditor's interest overrode notarial disinclination to identify people by means of addresses.

In contrast, only twenty-nine creditors, 3.5 percent of the total of 820, bothered to identify themselves by means of an address; the percentage is essentially the same as that found in the case of quittances. One of them was a woman named Adalays Arnosa, described as a *domina* in the records—a respectable woman, in other words, but not a noblewoman. Between 1356 and 1359, to judge by the extant records, Adalays had set herself up as a small-time moneylender, lending between three pounds and twenty florins. She used the services of the notary Jacme Aycart. In all six loans in which she represented herself before the notary, she insisted on including her address, the street of Francigena, after her name. One wonders whether she wanted to ensure that prospective clients would know where to find her. In a seventh loan, she was not present and was represented instead by a procurator, a cleric named Michael de Egerbio; in this act, her address was not given.[57] Chronologically speaking, in all but the first loan her debtors were also identified by means of address; all but one were agricultural laborers. Evidently this was someone who was exceptionally conscious of the importance of addressing.

As in the ancient Roman census or the fifteenth-century Florentine *catasto*, address, here, was used in a fiscal context. It was, however, not within the context of state fiscality. One of the origins for the use of the address in identity clauses lies in the burgeoning late medieval culture of debt, in particular, the possibility of credit-debt relationships between people, such as Christians and Jews or ladies and laborers, who were not otherwise bound by a relationship such as patronage, kinship, or friendship.

The ability to track down a debtor does not explain all the usages of address by members of the free Christian population. Addresses were never forced upon Jewish and noble debtors in notarized loans. It is perhaps understandable that a creditor named Peire Martin did not demand the address of the well-known knight, Johan Athos, to whom he loaned 170 florins in 1337. But familiarity does not necessarily explain why Inceff de Vidas, Clara de Isrrael, Jacop Mosse, Fossona de Oliolis, or any of the thirty-two Jews who borrowed money from Christians between 1337 and 1362 were not asked to provide addresses. The same is true of forty-eight Jews who bought things from Christians during the period and

---

[57] See ADBR 355E 9, fols. 5v–6r, 4 Apr. 1356; 355E 9, fol. 90r, 4 Dec. 1358; 355E 9, fol. 96r, 21 Dec. 1358; 355E 10, fols. 44r–v, 3 Sept. 1359; 355E 10, fol. 70r, 21 Oct. 1359; 355E 10, fols. 71r–v, 24 Oct. 1359; 355E 10, fol. 72v, 25 Oct. 1359.

Table 5.2. Comparison of cartographic categories used in notarial site clauses, notarial addresses, and vernacular addresses, 1337–62

	N	Streets (%)	Districts (%)	Vicinities (%)	Landmarks (%)
Notarial site clauses	932	58.3	16.6	16.4	8.7
Notarial addresses	249	59.0	19.7	13.3	8.0
Vernacular addresses	376	13.3	16.2	54.3	16.2

Sources: ADBR 2HD E7; 300E 6; 351E 2–5, 24, 641–45, 647; 355E 1–12, 34–36, 285, 290–293; 381E 38–44, 59–61, 64bis, 72–87, 393–94; 391E 11–18; AM 1 II 42, 44, 57–61.

subsequently acknowledged a debt. As these cases show, creditors or their notaries did not force the use of addresses on people who elsewhere show a reluctance to use them.

The last feature of notarial addresses to consider here is not how often addresses were used, but instead what sorts of addresses were preferred. In loans extant from the period between 1337 and 1362, there are 249 debtors whose domicile is indicated in the personal identity clause. The comparison of the cartographic categories used in site clauses and addresses is striking (Table 5.2). In both notarial site clauses and notarial addresses, the frequency of usage of streets, districts, vicinities, and landmarks is almost identical. Districts were used slightly more often in addresses, vicinities slightly less often. This is a reflection of the large number of poor suburban laborers who had recourse to loans. As we might have anticipated, *sixains* and islands are exceedingly rare in addresses. Islands were never used, and there is only one reference to a *sixain*, predictably the *sixain* of St. Jean. Compared, in turn, to vernacular addresses, the pattern of template usage in notarial addresses reveals even more clearly the scope of notarial translation from Provençal to Latin linguistic cartography.

To conclude, late medieval notaries were not significant agents in the gradual process by which addresses, in the practices of record-keeping bureaucracies, were attached to identities. As late as the mid-sixteenth century, notarial identity clauses included addresses only haphazardly, and the less systematic work I have done in notarial casebooks from later centuries suggests that the pattern continued. Where they did use addresses, they did so for two reasons: first, under the pressure of creditors interested in locating their debtors in space, and second, in a few cases where clients, especially clients of common status, seem to have insisted on using addresses as routine elements of their identities.

ADDRESSES IN SEIGNEURIAL RECORDS

The notarial disinterest in addressing stands in sharp contrast to the systematic use of addresses in documents of seigneurial origin recording the obligations of proprietors to pay yearly rents (Table 5.3). Here, we are faced with a relationship that was, like a loan, unequal. In the case of great landlords, like the bishop of Marseille, hundreds of individuals owed rents on lands and houses, and sophisticated records were required to keep track of them. Although property owners typically came voluntarily to the episcopal compound to pay their annual rents, episcopal officials had to know who had not paid and needed to be able to send a messenger to the property owner demanding his or her appearance. Hence, identity clauses in episcopal records include a great many addresses. A lengthy episcopal *levadou* or listing of owners of rural property from 1353, for example, reads almost like a telephone book, listing the addresses of 79.5 percent of the 192 men and women who appear in the record. Prepositions are frequent; in a sense, they tell the reader where to go to find the person in question. Uguo Niel, the record tells us, is at the Jewish fountain; Laurent de Lingris is in Corregaria; Douselina Seguiera is in the street of Pilis; Guilhem de Sant Peire is in the street of Guilhem Elie above the Spur; Giraut Alaman is next to St. Thomas. Only thirty-one individual proprietors were not given addresses. Most were great noblemen or important civic leaders such as Fulco Audebert, Johan de Jerusalem, and Raynaut de Sant Jacme. In light of the tendency of notarial records to record profession, trade, or status in identity clauses, it is particularly interesting that the official in charge of this episcopal register recorded the professions of only twelve proprietors. Noble status was *never* mentioned, despite the presence of numerous members of the nobility on the list. Clearly this list had a specific purpose, and that was to find people.

Officials working for the cathedral chapter, another major ecclesiastical landlord, were even more efficient than the bishop. Of 210 proprietors of vineyards, fields, and houses named in a register of rents from 1365, 87.1 percent were given addresses. Tellingly, when the notary Jacme Aycart was hired by the canons of the cathedral in 1350, in the wake of the Black Death, to perambulate the city collecting acknowledgments from thirty-nine property owners, the identity clauses he drew up included thirty-one addresses, or 79.5 percent. This is far more than was normal in notarial usage, for in 130 acknowledgments of rent owed from the notarial casebooks of the period, addresses were used 43 times or 33 percent of the time, similar to the frequency of addresses found in the

Table 5.3. Frequency of address use in seigneurial and confraternal records, 1335–65

Seigneur	N	Proprietors with addresses (%)	Proprietors without addresses (%)	Language of record	Source (ADBR) and date
St. Victor	65	89.2	10.8	Latin	1H 1146, 1357–59
Cathedral chapter	210	87.1	12.9	Latin	6G 485, 1365
Bishop	234	79.5	20.5	Latin	5G 115, 1353
Cathedral chapter (Jacme Aycart)	39	79.5	20.5	Latin	355E 3, fols. 8r–16r, 1350
Hospital of St. Esprit	129	69.0	31.0	Provençal	1HD B102, 1365
Confraternity of St. Jacques de Gallicia	560	67.1	32.9	Provençal	2HD E7, 1349–53
Marie de Jerusalem and Raymon del Olm	115	60.9	39.1	Provençal	1HD H3, 1348–60
Bernat Garnier	118	50.9	49.1	Provençal	4HD B1ter, 1335–45
Bernat Garnier	222	21.2	78.8	Latin	4HD B1quater, 1335–45
Jacme de Galbert	142	7.8	92.2	Latin	B538, 1348–49

case of loans. Jacme's use of addresses is particularly interesting because the acts are found in his regular casebook, that is to say, in a format where one might expect his normal customs to exert themselves. Evidently, Bernart Fulco, the priest who accompanied Jacme Aycart on this particular expedition, encouraged him to collect addresses from the proprietors.

The notary Peire Gamel, hired by the monastery of St. Victor to keep records concerning property under the dominion of the monastery, included addresses at a rate comparable to that of the cathedral canons in a register from the years 1357–1359. This register includes more than just acknowledgments of rent owed and therefore is not strictly comparable to the great rent registers kept by the episcopal curia and the cathedral canons, but if we consider only the acknowledgments and the acts of sale involving St. Victorine property in Peire's register, we find that 89.2 percent of the individuals with an obligation to the monastery were identified by means of an address. Two of the seven not given an address were Jews and a third was the well-known squire Carle Athos; all three were members of groups not normally identified by means of addresses.

Particularly interesting about the register kept by Peire Gamel are the emendations occasionally made to the addresses. One act originally

recording the fact that Berengiera Polhana lived at the Spur was changed
to read "now in the street of Prat Auquier," and in another act in which
the address was given simply as "Bernat Gasc" Peire added the marginal
notation: "the street of Bernat Gasc."[58] This suggests that he actually
verified addresses and changed them, if necessary, after the fact. In two
cases he noted that streets bore different names—"in the street of Bertran
Auriol, *alias* the street of the Fishmongery," and "in the burg of the
Dominicans in the Middle street, *alias* Madam Auriola's"—revealing a
similar concern for cartographic precision.[59] From the point of view of
translation of templates, perhaps the most interesting identity clause was
an address that originally read "near Malcohinat" which Peire later
emended to read "in the street of Malcohinat."[60]

Seigneurial interest in recording addresses also holds for other property lords, although addresses were recorded here somewhat less consistently. In a record of rents owed to the hospital of St. Esprit from 1365,
the rectors of the hospital, Bernat de Concas and Johan Johan, noted the
addresses of 69.0 percent of the proprietors. A similar record kept by
Marie de Jerusalem and her son Raymon del Olm from a decade earlier
did nearly as well. But perhaps the most interesting example of addressing in seigneurial records comes from the two records of rents kept by
the merchant, Bernat Garnier, already discussed in chapter three. In the
Provençal *levadou*, 50.9 percent of the proprietors are given addresses.
The Latin cartulary managed only 21.2 percent. The explanation for the
difference is enlightening, because the cartulary consists of a series of
abbreviated copies of acknowledgments of rent owed that were written
out in Bernat's register by the notary who had supervised the transaction.
The low figure of 22 percent, in short, reflects the customs of the diverse
group of public notaries who wrote the original acts and transcribed the
ensuing copies. The *levadou* itself was based on the acknowledgments kept
in the cartulary; in principle, therefore, it should have just as many
addresses as the cartulary. The fact that it has more than twice as many
addresses shows that Bernat, or someone in his pay, actively sought out
the information to add to the *levadou*.

Some of these seigneurial records suggest that the Black Death had an
effect on the seigneurial custom of addressing. The rent rolls listing rents
owed to Jacme de Galbert show a relatively low percentage of addresses,
partly because these rolls, like the Latin register kept by Bernat Garnier,

---

[58] ADBR H 1146, fol. 139v, 13 May 1359; fol. 8ov, 21 June 1358.
[59] Ibid., fol. 17r, 25 May 1357; fol. 74r, 27 May 1358.
[60] Ibid., fol. 170v, 31 Oct. 1359.

Table 5.4. Cartographic categories used in address clauses in seigneurial records, 1353–65

	N	Streets (%)	Districts (%)	Vicinities (%)	Landmarks (%)
Bishop	186	18.8	5.9	52.2	23.1
Cathedral	183	25.1	0	53.6	21.3

Sources: ADBR 5G 115 (1353), 6G 485 (1365).

were abbreviated copies of notarized acknowledgments of rent, and partly because the rolls list a significant number of noblemen who, as we have seen, seem to have been difficult to identify by means of domicile. The notary, Guilhem Tornator, began writing acknowledgments for Jacme de Galbert on rolls on 11 February 1348, shortly after the beginning of the plague, and none of the ninety-eight acknowledgments in the first two rolls includes an address. A year later, Jacme had his rents brought up to date, and hired the notary, Jacme Aycart, to do the job. The last of the three rolls includes acknowledgments made by forty-four proprietors. Eleven of these men and women were identified by means of addresses. It is probable that this just reflects the change in notary, but it is possible that the Black Death played a role too. The eleven new people, arriving into inheritances through torturous paths, may have been relatively unknown to the lord, and he may have encouraged the notary to provide more accurate identity clauses. Thus, in the same way that the plague brought about more property conveyances, thereby bringing about more cartographic conversations and hence cartographic standardization in notarial practice, rapid property turnover, also encouraged by the plague, may have diminished the possibilities of personalized lordship, where the lord was familiar with the proprietors who owed rent. Certainly the large, impersonal curias attached to the bishop and the cathedral were those whose records employed addresses most systematically.

A striking feature of the addresses found in seigneurial records is that they tend to follow the cartographic template typical of artisans and laborers which is in sharp contrast to the notarial habit of translating from the vernacular template to their Latin street template (Table 5.4; compare Table 5.2). As this illustrates, vernacular cartography had a tendency to push through the cartographic habits of the notaries or scribes responsible for seigneurial records. Episcopal and cathedral officials, it is true, used streets somewhat more often than did the confraternal scribe. Street usage is particularly marked in the case of the official responsible for keeping the register of rents owed to the canons of the cathedral from the year 1365. In this case, the higher usage of streets is partly related to the social characteristics of the proprietors, relatively few of whom lived in the suburbs where the use of burgs was common.

Yet we have a curious situation in the case of seigneurial records, for the addresses of individual proprietors were typically defined according to the norms of vernacular cartography, whereas the sites of the houses owned by the very same proprietors in the very same records were often defined by means of templates not favored in vernacular cartography, namely, the insular and street templates. Thus, seigneurial and vernacular cartographies jostled against one another in the very same act. Did no one notice the oddity? Possibly, but what is more to the point is that seigneurial officials clearly thought that the act of identifying unique individuals was entirely distinct from the act of identifying unique houses. As a result, they did not use the techniques developed for houses to identify people. When people gave their addresses and officials put those addresses to paper, neither of the parties concerned was interested in providing exact coordinates, the sort of exactitude, in the case of property, that was made possible by the dual action of site clauses and abutments. For curial officials, the vicinity where a person was known was enough; direct inquiries would lead them to the house of any defaulter. The fear that curial officials might do just that, namely publicize one's identity as a defaulter within one's vicinity, would also encourage would-be defaulters to find ways of fulfilling the obligation honorably. Knowledge of someone's vicinity was a potent weapon in the armory of seigneurial officials, since it gave them access to one of the well-springs of the reputation of the man or woman in question. Exactitude was irrelevant in this world of moral suasion.

The consistent use of address by property lords compares favorably with the usage of addresses found in Marseille's courts, since in the records of Marseille's civil courts, addresses are rare, about as uncommon as they are in notarial records. This may reflect the simple fact that trial transcripts were kept by notaries, but also reflects the fact that other features of identity, notably marital status and parentage, gave rise to significant legal rights in the property or estates of other people, and hence were noted more often in identity clauses in records of the civil courts. There are several extant financial registers of the *clavaires* of Marseille that listed fines generated by the criminal court. As seen earlier in the chapter, these records were very spare, and rarely include addresses; their purpose was financial, not criminological.[61] In its own records, the court of inquest

[61] See ADBR, series B, 1940, 1943, 1944, 1945, 1947, 1949, 1950, and 1953.

itself noted the addresses of defendants with a consistency approaching that of the great property lords. A record from the year 1380 included the names of one hundred and five individuals accused of a variety of crimes, and seventy-eight or 74 percent were given addresses. In the case of a shoemaker named Cliquinus, the record noted that he was a vagabond.[62] But it is difficult to find any obvious trend toward greater addressing; among other things, a similar record of the activities of the criminal court from 1465 includes no addresses whatsoever.[63]

The habit of attaching identity to a place was neither particular to Marseille nor especially new. In Icelandic sagas, chieftains and householders are identified not only by means of genealogy but also by residence. The earliest noble families of the former Carolingian lands took toponyms from their castles. In medieval and Renaissance Italian cities, nobles and patricians habitually identified themselves with their palaces and compounds. What is striking about Marseille is that this habit was not confined to the landholding aristocracy. Moreover, in Marseille the habit could not have trickled down from aristocratic identity practices, since Marseille's noble families and great merchants rarely identified themselves in bureaucratic records by means of house or street. This is not to suggest that Marseille's nobility was primitive or its non-noble population advanced. Identity constructs vary according to their contexts; what serves as an identity in narrative sources will not necessarily translate to bureaucratic contexts. Among noble families in Marseille, identity was established primarily by means of genealogy, not unlike the genealogical form that dominated strategies of identifying islands and houses in seigneurial record-keeping practices. Artisans were not genealogically oriented. They identified themselves as being members of trade groups. They did not use addresses often, but when they did, they often chose to identify themselves with vicinities that had social and moral purposes.

The will to address in fourteenth-century Marseille came most notably from creditors and seigneurial officials eager to locate their debtors in space. Here, the context was decidely coercive; one can say that creditors were attaching addresses to individuals. In the case of some seigneurial records, this practice was so consistent as to have constituted a norm of record-keeping practices. Recording an address not only allowed officials to know where to find a given defaulter; it would also have allowed officials to tap into the vicinity, the roots of a person's reputation.

[62] ADBR 3B 96, fol. 10r.
[63] ADBR 3B 172.

Modern urban addresses, complete with street names and numbers, did not begin to develop until the eighteenth century, and hence lie outside the framework of this study. According to the arguments made here, they emerged at the confluence of two cartographic streams. On the one hand, we have the growing tendency of seigneurial officials and other creditors to affix addresses to identities. These addresses were often framed as vicinities, a moral or social construct more than a rational description of space. On the other hand, we have notaries who were fostering the rationalized, street-based cartography that became the urban norm in the seventeenth or eighteenth century. As the two streams merged, it became possible to imagine addresses defined with the exactitude once reserved for property sites.

Epilogue

The twentieth century has been the century of the nation-state, and has therefore seen the flowering of a historiographical literature devoted to exploring the origins of the state. By any definition, modern states always look better on paper than medieval states. Medieval historians, especially in the United States, once bridled at the suggestion of medieval political infantilism, and it has long been common for medieval historians of politics and power to emphasize the rational and calculating vision of medieval sovereign powers who were Machiavellian *avant la lettre*.[1] The literature offers us the image of medieval states marked, from the twelfth century onward, by a growing interest in the rationalization and standardization of fiscal policy, justice, military technology, military organization, and record keeping, relatively unencumbered by the pious platitudes of Christian political philosophy.[2] This political transformation, in turn, was deemed to be the political expression of what is often called the Renaissance of the twelfth century.[3]

There is, of course, much that is mythical in the very idea of a medieval state, and good reason for avoiding the overly optimistic interpretations

---

[1] This trend is epitomized by Joseph R. Strayer, *On the Medieval Origins of the Modern State* (Princeton, 1970).

[2] Some recent examples include John Baldwin, *The Government of Philip Augustus: Foundations of French Royal Power in the Middle Ages* (Berkeley, 1986); David Abulafia, *Frederic II: A Medieval Emperor* (London, 1988). Also relevant is Ernst Cassirer, *The Myth of the State* (New Haven, 1946).

[3] A term introduced by Charles Homer Haskins in his *The Renaissance of the Twelfth Century* (Cambridge, Mass., 1927).

of earlier scholars.[4] The debate has been joined and promises to be a fruit-ful one. But for all the disagreements, both sides share a critical episte-mological assumption that I have tried to challenge in this book. They share the idea that states do all the thinking.[5]

What I mean by this is fairly simple. States, or more accurately the agents of state institutions and bureaucracies, necessarily use categories of understanding to help guide policy. To have a policy on vagrants, one must put some thought into what a vagrant is, and the category itself is defined, and refined, as the policy develops. The same is true for laws that apply to categories of persons such as peasants or types of actions such as murder or rape. Similar principles apply in less obvious areas, such as fiscality, for it would have been impossible for a medieval exchequer to assess royal revenue without some fundamental unit of currency according to which all forms of revenue could be measured, including taxes collected in different forms of currency as well as non-currency items such as chickens, barrels of wine, cheeses, grain, wax, and so on. We know, of course, that a lot of the thinking gets done by the institu-tions of civil society and not by the state, so merchants are as important as finance ministers in lending definition to units of currency. It is guilds of weavers and drapers who define the measure and make of fabric; it is physicians who define lepers; it is jurists not responsible to any state who define many of the categories of the *ius commune*. Whatever these excep-tions, however, it is seductively easy to assume that states create, or at the very least promote, the development of basic categories of thought and apprehension. When we try to assess a given government's degree of impersonality in office holding or rationality of practice, very often what we assess, however unconsciously, is the degree to which governmental records reveal the existence of formal identity templates and identity categories. We often think of modern states as imposing these identity templates on an intellectually passive population. This is how a citizenry was shaped.

---

[4] Professor Bisson, for example, has been a sharp critic of the idea of a medieval state in his recent work; he stresses instead the patrimonial or seigneurial nature of medieval power. See, inter alia, T. N. Bisson, "The 'Feudal Revolution'," *Past and Present* 142 (1994): 6–42; idem, ed., *Cultures of Power: Lordship, Status, and Process in Twelfth-Century Europe* (Philadel-phia, 1995).

[5] Fredric L. Cheyette has discussed the very close association between categories of thought and the state in "The Invention of the State," in *Essays in Medieval Civilization*, ed. Bede K. Lackner and Kenneth R. Philp (Austin, Texas, 1978), 143–78. Foremost among Cheyette's aims is to ascribe to the eleventh-century papacy the intellectual development of a cate-gorical distinction between office and person, an idea, he claims, that was unthinkable before.

Consider the example of names. Speaking of the invention of permanent, inherited patronyms, James C. Scott has remarked that:

> In almost every case it was a state project, designed to allow officials to iden-
> tify, unambiguously, the majority of its citizens. When successful, it went far
> to create a legible people. Tax and tithe rolls, property rolls, conscription
> lists, censuses, and property deeds recognized in law were inconceivable
> without some means of fixing an individual's identity and linking him or
> her to a kin group.[6]

Although Joseph Strayer did not discuss the subject in so many words,
there is little here that would have offended his argument about the
medieval origins of the modern state. Yet how accurate is the assumption
that states do all the thinking? Scott's observations may well be accurate
in the context of modern colonialism and imperialism. But unless we are
willing to accord far more agency to medieval states and medieval bureaucracies than we have hitherto been accustomed, it cannot be true of the
central and later middle ages, the period in which most European surnames were invented. As I suggested in the previous chapter, notarial and
scribal culture played a dominant role in the invention of the surname,
and notarial culture, especially in Mediterranean Europe, was at the
outset only loosely associated with state interests. As argued earlier, moreover, mnemotechnics were a significant feature of the notarial act of
identification, and thus the surname, although useful, was in no way
required by some given bureaucratic template. Far from being incon-
ceivable, tax and tithe rolls, property deeds, and the other records listed
by Scott could operate quite efficiently as long as the work of
identification and categorization was entrusted to memory. The anachro-
nism here, I think, comes from assuming that states have always done all
the thinking, and that "thinking," namely the act of categorizing, classifying, and simplifying, cannot be done by a largely autonomous and
decentralized record-keeping bureaucracy such as the late medieval
notariate.

Closer to the preoccupations of this book, it has become de rigueur to
speak of state interest in and control over early modern developments in
mapping.[7] Cartography seems so very much like the sort of thing that
early modern states would have been eager to promote, an act of political patronage that in turn created standard and universalizing categories

---

[6] James C. Scott, *Seeing Like a State: How Certain Schemes to Improve the Human Condition Have
Failed* (New Haven, 1998), 65.
[7] See above, pp. 3–6.

of spatial apprehension. This literature has been an improvement over an older literature that explained the cartographic revolution as a manifestation of a scientific revolution. All the same, the state and its self-interest have become a new teleology, replacing the progressivism of nineteenth-century historiography but doing little to challenge a theory that locates the motive force for change in some monolithic principle. I do not doubt that early modern states or their agents developed an interest in thinking about space and did so in the interests of power. But that is not why this kind of thinking first emerged, if the arguments of this book have any merit. We need a paradigm of historical change broad enough to take some loosely defined "state interest" like this and disassemble it into its myriad components and its genealogies, for as things stand now it is not easy to describe the historical emergence of categories of thought without assigning interest and agency to a self-conscious state. We sometimes assume too readily that to explain the evolution of categories of thought, we need only see how these categories function in a later society. If categories of space were developing in the later middle ages, so the argument would go, they must have been developing for the purposes to which they were eventually put.

In this book I have challenged this teleology by locating agency not in a faceless and monolithic state but rather in an array of smaller, more humanized sets of interests. I have argued that two relatively small and cohesive interest groups in Marseille, namely the public notaries and also, to some extent, seigneurial officials, played a role in important changes in cartographic or spatial discourse, particularly in the ways in which they mapped out possession and identity. I have also argued that their interest cannot be associated with state-level interests. Public notaries in late medieval Marseille created a standard cartographic language of streets, and by the sixteenth century the language had caught on in the practices of other agents of record and had become, in effect, the official cartography of Marseille. This took place well before the development of any readily apparent state interest in urban cartography. By the eighteenth century, we have royal ordinances enjoining the numbering of houses, and this is obviously a component of a rational-legal description of urban space, but the numbering of houses could only have been conceivable after streets became the basic unit of cartographic awareness, a development that took place in the preceding centuries and was promoted by public notaries. In their rent registers, in turn, late medieval seigneurial officials were involved in creating an identity template of which the major component was the address of the proprietor. This development, certainly not unique to Marseille, shows how categories of identity could develop outside the aegis of the state. A state like that of fifteenth-century Florence could also attach individuals to domiciles—most notably in the records of the Catasto of 1427–1430—but this does not diminish the fact that domicile as a category of identity also had autonomous and possibly earlier origins. Streets and official domiciles later became important elements of modern bureaucratic identity templates, contributing to the political regime of modern nation-states by crafting identities that are cellular and modular. But they first emerged as categories of thought within a late medieval culture of record that should be a proper object of historical research.

What was this culture? Why did it change or standardize the categories of cartographic awareness? To reiterate a point made throughout the book, the public notariate, continuously developing in Europe from the middle or late twelfth century onward, was not an organ of a particular state. The legal genealogy was rooted in the international *ius commune*, and notarial culture itself was heavily shaped by the usage to which it was put by consumers of the law. For these reasons, the notariate, a curious form of autonomous, decentralized bureaucracy, is scarcely acknowledged in political histories of medieval bureaucracy and administration. In addition to being legal agents or brokers, governing people's access to legitimate forms of force increasingly controlled by the courts of law, notaries were also agents of knowledge, authoritative depositories of facts regarding identity and property not by virtue of their legal capacity so much as by their capacity to categorize, archive, and remember facts and to construe identities. As an archival bureaucracy, the public notariate created its own categories of intellectual apprehension, using the legal categories defined by the jurists of the *ius commune* wherever possible but inventing categories where guidance was not offered. Among other things, they were or became cartographers, as repeated cartographic conversations in legal contexts made them authoritative knowers of the city plan. This development may have been predictable in the twelfth century but was hardly foreordained. As suggested in chapter two, we cannot really know why notaries in Marseille promoted the cartographic category of the street, something that was at odds with the cartographies used by other agents of record, such as the island, the vicinity, and the landmark, as discussed in chapters three and four. There is nothing I have seen in the records from Marseille to suggest that notaries were conscious of what they did. It may be that they were troubled by the imprecise nature of alternative cartographies and sought to reformulate the science of mapping, but then we are left with the problem of explaining how they came to be troubled by it. It seems easier to suggest that the notaries, as largely unselfconscious cartographers, were casually inventing a language for something that people hadn't thought about a great deal before, namely space. In much the same way we can imagine a neolithic population, drifting northward, inventing a language for snow, or artisans, after the French Revolution, finding it necessary to invent a new language for labor.[8]

Once the language became relatively universal, spreading outside notarial circles by virtue of the repeated conversations between notaries and their clients, the implications of this new mapping of space would develop on their own. Among other things, a cartographic language developed for use in property conveyances became increasingly available for the mapping of citizen identities, because an address, once thinkable, becomes convenient for denoting precise individuals—and, of course, for finding them in the event of breach of contract. Hence, creditors have an interest in this developing language, as discussed in chapter five. State bureaucracies or parastatal institutions (like post offices) eventually recognized the value of the address, thus borrowing and giving increasingly more purpose and meaning to a practice that nonetheless, in its origins, had no internal logic, no DNA code guiding its future development.

If there is merit to this argument, then the later middle ages takes on a political and cultural significance it has never had before. The implication of the use of the term "Renaissance," after all, is that what happened in the fifteenth and sixteenth centuries was a break from the immediate past. If few historians, and in particular medieval historians, are deluded by the self-serving rhetoric of Renaissance commentators, that does not mean we have studied the later middle ages as a period of cultural production.[9] We sometimes assume that whatever happened in the later middle ages was laid down in the twelfth century and followed predictable lines of development, whereas the sixteenth century saw the introduction of new social and political forces. In the realm of linguistic cartography, we see that this interpretation of the later middle ages as an appendage of the twelfth century does not work. Little in twelfth-century notarial culture could have been used to predict how notaries in Marseille and elsewhere became authoritative cartographers. At a stroke, the later middle ages becomes a fascinating site of cultural production, where existing institutions developed new interests, priorities, and languages for

[8] William H. Sewell, Jr., *Work and Revolution in France: The Language of Labor from the Old Regime to 1848* (Cambridge, 1980).
[9] The famous exception is Johan Huizinga, *The Autumn of the Middle Ages*, trans. Rodney J. Payton and Ulrich Mammitzsch (Chicago, 1996).

highly contingent and unpredictable reasons. These languages, as they emerged, took on their own lives and were taken up by other knowledge groups, including state-level bureaucracies. By studying how languages and categories evolved, we have a way to think of the later middle ages as an element in a seamless historical development linking the middle ages to the sixteenth century and centuries beyond. This kind of linkage gives us reason to hope for, and push for, a medieval historiography that will once again be firmly integrated into the ways in which we think about the development of state and society in western Europe.

APPENDIX 1

Lexical Terms Used in the Register of the Confraternity of St. Jacques de Gallicia, by Category

Artisanal and retail vicinities

Frucharia (Fruitery)	23
Pescaria (Fishmongery)	15
Frenaria (Jewelery)	7
Triparia (Tripery)	6
Auruvellaria (Goldsmithery)	5
Curataria (Cobblery)	5
Fustaria (Carpentery)	5
Sabataria de San Jacme (Shoemakery of St. Jacques)	5
Mazel Veilh (Old Market)	4
Pellissaria (Furriery)	4
Botaria (Coopery)	3
Draparia (Drapery)	3
Draparia Sobeyrana (Upper Drapery)	3
Patinaria (Slipperery)	3
Blancaria (Tannery)	2
Cordelaria (Cordery)	2
Coutellaria (Cutlery)	2
Pelisaria Estrecha (Narrow Furriery)	2
Pelisaria Larga (Wide Furriery)	2
Annonaria (Grain Market)	1
Bossaria (Bucklery)	1
Cambis (Change)	1
Gran Mazel (Grand Market)	1

Some spellings have been normalized and various spellings of the same toponym have been grouped under a single term. No translations have been given for ecclesiastical institutions or obscure Provençal toponyms.

[229]

Grolaria (Shoemakery)   1
Mercat (Market)   1
Sabataria (Shoemakery)   1
Sabataria del Temple (Shoemakery of the Temple)   1
Veyraria (Glaziery)   1
                                                         Total   110

Landmark vicinities

Esperon (Spur)   18
Escas (Pilings)   12
Font Jueva (Jewish fountain)   12
Corregaria   11
Colla (Hill)   10
Prat d'Auquier   8
Cort de la Marquesa (Court of the Marquesa)   3
Crotas (Arches)   3
Cassayria   2
Forn dan Prodome (Gentleman's oven)   2
Lansaria (Lancery)   2
Malcohinat   2
Peyra que Raja (Roiling Stone)   2
Tonnieu (Tholoneum)   2
Bella Taula (Pretty Table)   1
Enquant (Auction)   1
Escalas (Steps)   1
Labeurador   1
Lauret   1
                                                         Total   94

Landmarks

Frayres Menos (Franciscans)   4
Accolas (Notre Dame des Accoules)   3
Costa Peire Arman (next to Peire Arman)   3
Prezicadors (Dominicans)   3
Riba (Quay)   3
St. Augustin   3
St. Catharina   3
St. Martin   3
Temple   3
Davant lostal de sen Peire Austria (before the house of Peire Austria)   2
Forn de la Botaria (oven of the Coopery)   2
Forn de la Curataria (oven of the Cobblery)   2
Forn de Riqui Novas (oven of Riqui Novas)   2
Forn del Esperon (oven of the Spur)   2
Frache   2

Syon	2
St. Esprit	2
A mazon essen Jacme Donadieu (in the house of Sir Jacme Donadieu)	1
Costa Folco Audebert (next to Folco Audebert)	1
Costa Johan Sancho desot la Pescaria (next to Johan Sancho below the Fishmongery)	1
Costa Raolin (next to Raolin)	1
Davant Guilhem Estaca peyrier (before Guilhem Estaca, mason)	1
Davant Peire Ameli (before Peire Amiel)	1
Denant Guilhem Tomas a Lesperon (before Guilhem Tomas, at the Spur)	1
Denant Marques Malet (before Marques Malet)	1
En l'ostal de Johan de Limojes (in the house of Johan de Limojes)	1
Forn de Cavalhon (oven of Cavalhon)	1
Palays (Palace)	1
Porta Gallega (Gallican gate)	1
St. Clare	1
St. Jaume	1
St. Laurens	1
St. Loys	1
St. Paul	1
Total	61

Streets

C. Negrelli, c. Negrel (Negrel street)	13
C. de l'Almorna (street of the Almoner)	3
C. de Preziquados (street of the Dominicans)	3
C. St. Augustine, c. de Sant Agostin (street of St. Augustine)	3
C. St. Martin (street of St. Martin)	3
C. de Jaret (street of Jaret)	2
C. de las Marquezas (street of the Marquesas)	2
C. de Robaut (street of Robaut)	2
C. dels Fabres, c. Fabrorum (street of the Smiths)	2
C. Cavalhoni (street of Cavalhon)	1
C. de Frayres Menos (street of the Franciscans)	1
C. de Prat Auquier (street of Prat d'Auquier)	1
C. de Sant Martin (street of St. Martin)	1
C. de Sion (street of Syon)	1
C. del Perier (street of Perier)	1
C. del Portal del Laurier (street of the gate of Lauret)	1
C. dels Bochas (street of the Bochas)	1
C. den Castilhon (street of Castilhon)	1
C. den Garrian (street of the Garrian)	1
C. den Raymon Rasquas (street of Raymon Rasquas)	1
C. Franseza (street of Francigena)	1
C. Guilhem Folco (street of Guilhem Folco)	1
C. Guilhem Imbert (street of Guilhem Imbert)	1

C. Jaume Cancel (street of Jaume Cancel)   1
C. Oleriorum (street of Oliers)   1
C. sen Antoni Tozesco (street of Antoni Toesco)   1
                                                    Total   50

Districts (except burgs)

Cavalhon   23
St. Johan   6
Anonaria Sobeyrana (Upper Grain Market)   2
Rocabarbola   2
Juzataria (Jewry)   1
                                                    Total   34

Burgs

Burg Oleriorum (burg of Oliers)   8
Burg Moreriorum (burg of Moriers)   7
Burg de Sion (burg of Syon)   6
Burg St. Augustin (burg of St. Augustine)   3
Burg dels Preziquados (burg of the Dominicans)   2
Burg de Jaret (burg of Jaret)   1
                                                    Total   27

                                                    Total   376

source: ADBR 2HD E7

The Prosopographical Index

The prosopographical index is a profile index of all the people who appeared in all the records I used from the period from 1337 to 1362. The database, which has slightly more than 17,500 individual files, is the result of a labor intensive linkage of records. A few typical entries are shown below.

name:	Guillelmus Cabrerii
alternative spellings:	Cabre
profession:	laborer
origin:	Aubagne; now citizen and resident
city:	lower city
quarter:	Callada
neighborhood or street:	Curataria
source:	1359 (1HD H3); 11JA; 12JA; 3B60 (1358) 10; 1GJ; 12JA (1362) 21, 22
comment:	owns 1/3 *ostal* with Uguo Rogier and Antoni Martini behind Curataria

name:	Antoni Catalani
alternative spellings:	none
profession:	merchant
origin:	citizen
city:	lower city
quarter:	St. Jacobi
neighborhood or street:	carreria Corregarie
source:	BB 23 (1361); 1360a; 2JA; 6JA (1353) 86; 9JA (1358) 94, 116, 125; 10JA (1359) 16, 37, 102, 127; 12JA (1362) 42, 96; 6G (1365); 1341 (4HD

Biter); 3PG; 3PG (1348) 16; 4PG; 6PG; 7PG (1359) 201; 1PlG (1337) 84, 153; 3B 57 (1355) 25; 3B 48 (1351) 23; 3B 41 (1340) 11; 3B 811 (1352) 7; 3B 42 (1341) 2

comment: wife Blanca; godkin of Stephanus de Sancto Paulo; acting for Bertran de Massilia, of Oliolis, in 9JA; just sold house in Corregaria in 1341; son of Macellus Catalani [died 1337]; actor for noble Paulus de Villanova, dominus de Vensa; witness for Bertran Gontardi

name: Alasacia Columberia
alternative spelling: also married name de Rupeforti
profession: notary (husband)
origin: citizen
city: lower city
quarter: Acuis
neighborhood or street: *supra* Fontem de Acuis
source: 6G (1365); 5G116 (1365); 6PA (1354) 2; 12JA (1362) 132, 159
comment: wife/widow of Rostagnus [died 1362]; aunt of Columberia de Rupeforti, wife of Antoni de Rupeforti

The record is headed by the Latin name for the person in question, followed by other Latin or Provençal spellings. Women present a special difficulty since their surnames changed at marriage. Where known, I used maiden name and noted the married name in the "alternative spelling" cell. In the case of Alazais Columberia above, I did not know her maiden name, and thus filed her under the surname of her husband. One record indicated that she was also once married to a person by the surname of "de Rupeforti," so I noted that fact in the "alternative spelling" cell.

After the name comes information on status, such as trade, profession, "noble," or "Jew." For women, I used the status of their fathers and/or husbands, where known. Then comes information on residence. In the file of the laborer Guilhem Cabre (*Guillelmus Cabrerii*) above, we know that he was once a citizen of the small town of Aubagne a few miles southwest of Marseille, although he had become a full citizen and resident of Marseille. What I judged to be his place of residence was determined by his ownership of part of a house (*ostal, hospicium*) in the neighborhood known as Curataria, the Cobblery. Knowing that the Cobblery was located in the *sixain* of Callada in the lower city, I filled in the remainder of the information in his record. In many cases, of course, I did not know where boundary lines of the *sixains* of the lower city fell; in such cases, records like that of Antoni Catalan allowed me to determine where boundaries could be drawn. In his record, we know that he owned one or more houses on the Corregaria, a street near the border between the *sixains* of St. Jacques and Draparia. We also know from the register of deliberations of the city council that he was a council member for the *sixain* of St. Jacques in 1361, and we know from the taille of 1360–1361 that he owned at least one property in the *sixain* of St. Jacques. From this I drew the conclusion that Corregaria was located in the *sixain* of St. Jacques, a conclusion amply confirmed in other sources.

The "source" entry lists abbreviations of all the sources in which the person appeared. The year was included in the abbreviation so as to make it relatively easy to determine when a given person was active in the records. The last cell, "comments," includes any important miscellaneous information, especially kinship.

I used the following strategy to link the many thousands of individual records. First, in taking notes on notarial and judicial documents using ordinary word-processing software (at the time, WordPerfect version 4.2), I included indexing codes in my records. When my note-taking was complete, these codes allowed me to run a simple macro that picked out from each record the names of the people that I wanted indexed, all the information given about them in the record, and an abbreviation of the record's name and number. What resulted was a crude list of thousands of names and bits of information in a non-standard format; each item, then, I transferred (again using macros, where possible) one by one into my standard record format. For the rent registers, the process was less labor-intensive, because the information in these registers is much more regular than the information in notarial casebooks and court registers. This regularity allowed me to keep my original notes in a standard format, considerably simplifying the indexing process.

What resulted, then, were extensive lists of individuals and the sources in which each individual appeared. I began compressing the records that resulted from my indexing of notarial casebooks notary by notary, on the assumption that notaries frequently dealt with the same clientele, therefore making certain identifications more likely. After sorting by surname (and later by first name), I made linkages according to several principles. Identical name, of course, was the most important principle, although I used this uncritically only in cases where either the surname or the first name was uncommon. Taking the first case above as an example, both Guillelmus and Cabrerii were relatively common names, so in this case I searched for a better indication of identification before making any linkages. The surviving records indeed show that there was a butcher by the name of Guillelmus Cabrerii (spelled *Caprerii*, which was essentially the same surname), and I distinguished his records carefully from those of the laborer. A third person named Guilhem Cabrier shows up in the *taille* of 1360–1361 as a resident of the *sixain* of Acuis. In all likelihood this third Guilhem was in fact the butcher, since the butchers had a market in the *sixain* of Acuis and many of them lived around it, although not being certain I did not make this linkage. In cases like this where no unimpeachable identifiers could be found to distinguish one person from another, I usually gathered the information together in a single file and labeled it miscellaneous, for example, "Guillelmus Cabrerii (miscellaneous)." If this person should later turn out to be someone I might have an interest in, his history and identity could be more easily reconstructed.

At an early stage in this process, I was careful to keep the information in compressed records segregated by source and information so that, if the identification turned out later to be faulty, the record could easily be disassembled. Where the identification bordered on the questionable, I also included a note to myself in the record indicating that there were grounds for doubting the trustworthiness of the linkage.

When I had compressed all the records from notarial sources in this way, I assembled all the records—notarial, rent, judicial, fiscal, miscellaneous—in a master file of 26,000 or more records that was then sorted and, continuing further, reduced by about a third to the final total of 17,500. When that project was more or less complete I eliminated by hand any redundant information and finished the process of transferring information, notably residential information, to individual cells.

The making of linkages is a subjective process. As I suggested above, one cannot link every record on the basis of name alone. Nevertheless, if one is too conscientious and never follows a hunch, a great percentage of very reasonable linkages would never get made. Without question the database includes some errors, although my subsequent experience in working with it suggests that the most common of these errors is the erroneous compressing of the files of a father and a son of the same name. On the whole, I have tried to err on the side of conservatism.

Bibliography

Abulafia, David. *Frederic II: A Medieval Emperor.* London: Penguin, 1988.

Abu-Lughod, Janet L. "The Islamic City—Historic Myth, Islamic Essence, and Contemporary Relevance." *International Journal of Middle East Studies* 19 (1987): 155–76.

Adams, Thomas McStay. *Bureaucrats and Beggars: French Social Policy in the Age of the Enlightenment.* Oxford: Oxford University Press, 1990.

Akerman, James R. "The Structuring of Political Territory in Early Printed Atlases." *Imago Mundi* 47 (1995): 138–54.

Alliès, Paul. *L'invention du territoire.* Grenoble: Presses Universitaires de Grenoble, 1980.

Amargier, Paul. "Mouvements populaires et confrérie du Saint-Esprit à Marseille au XIIIᵉ siècle." In *La religion populaire en Languedoc du XIIIᵉ siècle à la moitié du XIVᵉ.* Vol. 11 of *Cahiers de Fanjeaux,* 305–19. Toulouse: Privat, 1976.

Amelang, James S. "People of the Ribera: Popular Politics and Neighborhood Identity in Early Modern Barcelona." In *Culture and Identity in Early Modern Europe (1500–1800),* edited by Barbara B. Diefendorf and Carla Hesse, 119–37. Ann Arbor: University of Michigan Press, 1993.

Anderson, Benedict. "Census, Map, Museum." In *Becoming National: A Reader,* edited by Geoff Eley and Ronald Grigor Suny, 243–48. New York: Oxford University Press, 1996.

——. *Imagined Communities: Reflections on the Origin and Spread of Nationalism.* London: Verso, 1983.

Aubenas, R. *Étude sur le notariat provençal au moyen âge et sous l'Ancien Régime.* Aix-en-Provence: Éditions du Feu, 1931.

Bagnall, Roger S., and Bruce W. Frier. *The Demography of Roman Egypt.* Cambridge: Cambridge University Press, 1994.

Baldwin, John W. *The Government of Philip Augustus: Foundations of French Royal Power in the Middle Ages.* Berkeley: University of California Press, 1986.

Baratier, Édouard. *La démographie provençale du XIIIe au XVIe siècle.* Paris: S.E.V.P.E.N., 1961.

——. *Histoire de Marseille.* Toulouse: Privat, 1973.

Baratier, Édouard, and Félix Reynaud. *De 1291 à 1480.* Vol. 2 of *Histoire du commerce de Marseille.* Chambre de Commerce de Marseille, edited by Gaston Rambert. Paris: Plon, 1951.

Barkey, Karen. *Bandits and Bureaucrats: The Ottoman Route to State Centralization.* Ithaca: Cornell University Press, 1994.

Barthes, Roland. *Empire of Signs.* Translated by Richard Howard. New York: Hill and Wang, 1982.

Bartlett, Robert. *The Making of Europe: Conquest, Colonization, and Cultural Change, 950–1350.* Princeton: Princeton University Press, 1993.

Bauman, Richard, and Joel Sherzer, eds. *Explorations in the Ethnography of Speaking,* 2d ed. Cambridge: Cambridge University Press, 1989 [1974].

Bautier, Robert-Henri. "Les foires de Champagne: recherches sur une évolution historique." In *La Foire.* Vol. 5 of *Recueils de la société Jean Bodin,* 97–147. Brussels: Éditions de la Librairie Encyclopédique, 1953.

Bautier, Robert-Henri, and Janine Sornay. *Les sources de l'histoire économique et sociale du moyen âge.* 3 vols. Paris: Centre National de la Recherche Scientifique, 1968–84.

Beaune, Colette. *The Birth of an Ideology: Myths and Symbols of Nation in Late-Medieval France.* Translated by Susan Ross Huston. Edited by Fredric L. Cheyette. Berkeley: University of California Press, 1991.

Bellomo, Manlio. *The Common Legal Past of Europe, 1000–1800.* Translated by Lydia G. Cochrane. Washington D.C.: Catholic University of America Press, 1995.

Bencivenne. *Ars notarie.* Edited by Giovanni Bronzino. Bologna: Zanichelli, 1965.

Benjamin of Tudela. *The Itinerary of Benjamin of Tudela: Travels in the Middle Ages.* Edited by Michael A. Signer. Malibu, Calif.: Joseph Simon and Pangloss Press, 1983.

Benoit, Fernand, et al. *Monographies communales, Marseille-Aix-Arles.* Vol. 14, part 3 of *Les Bouches-du-Rhône. Encyclopédie Départementale.* Edited by Paul Masson. Marseille: Archives Départementales des Bouches-du-Rhône, 1935.

Berger, Peter L., and Thomas Luckmann. *The Social Construction of Reality: A Treatise in the Sociology of Knowledge.* Garden City, N.Y.: Doubleday, 1966.

Berlière, Jean-Marc. *Le monde des polices en France: XIXe–XXe siècles.* Brussels: Éditions Complexe, 1996.

Berlow, Rosalind Kent. "The Sailing of the 'Saint Esprit'." *Journal of Economic History* 39 (1979): 345–62.

Bernardi, Philippe. "Métiers du bâtiment et techniques de construction à Aix-en-Provence à la fin de l'époque gothique (1400–1550)." Thèse, Université de Provence Aix-Marseille I, 1990.

Bertillon, Alphonse. *Identification anthropométrique.* 2 vols. Melun, 1893.

Biddick, Kathleen. "Paper Jews: Inscription/Ethnicity/Ethnography." *Art Bulletin* 78 (1996): 594–99.

Bisson, T. N. "The 'Feudal Revolution'." *Past and Present* 142 (1994): 6–42.

——, ed. *Cultures of Power: Lordship, Status, and Process in Twelfth-Century Europe.* Philadelphia: University of Pennsylvania Press, 1995.

Black, Jeremy. *Maps and Politics*. London: Reaktion Books Ltd, 1997.

Blancard, Louis. *Documents inédits sur le commerce de Marseille au moyen âge, édités intégralement ou analysés*. 2 vols. Geneva: Mégariotis Reprints, 1978 [1884].

Blès, Adrien. *Dictionnaire historique des rues de Marseille*. Marseille: Éditions Jeanne Laffitte, 1989.

Bloom, Paul, et al., eds. *Language and Space*. Cambridge, Mass.: MIT Press, 1996.

Boltanski, Luc. *Les cadres. La formation d'un groupe social*. Paris: Minuit, 1982.

Bouiron, Marc. "Le fond du Vieux-Port à Marseille, des marécages à la place Général-de-Gaulle." *Méditerranée* 3 (1995): 65–68.

Bourin, Monique, and Pascal Chareille, eds. *Désignation et anthroponymie des femmes. Méthodes statistiques pour l'anthroponymie*. Part 2 of *Persistances du nom unique*. Vol. 2 of *Genèse médiévale de l'anthroponymie moderne*. Tours: Publications de l'Université de Tours, 1992.

Bourrilly, Victor-L. *Essai sur l'histoire politique de Marseille des origines à 1264*. Aix-en-Provence: A. Dragon, 1925.

Boutier, Jean, Alain Dewerpe, and Daniel Nordman. *Un tour de France royal: le voyage de Charles IX (1564–1566)*. Paris: Éditions Aubier Montaigne, 1984.

Brezzi, Paolo, and Egmont Lee, eds. *Sources of Social History: Private Acts of the Late Middle Ages*. Toronto: Pontifical Institute of Mediaeval Studies, 1980.

Broise, Henri. "Les maisons d'habitation à Rome aux XVe et XVIe siècles." In *D'une ville à l'autre: structures matérielles et organisation de l'espace dans les villes européennes (XIIIe–XVIe siècle)*, edited by Jean-Claude Maire Vigueur, 609–29. Rome: École Française de Rome, 1989.

Broise, Henri, and Jean-Claude Maire Vigueur. "Strutture famigliari, spazio domestico e architettura civile a Roma alla fine del medioevo." In *Storia dell'arte italiana*, 99–160. Turin: Einaudi, 1983.

Brown, Lloyd A. *The Story of Maps*. Boston: Little, Brown, 1949.

Buisseret, David, ed. *Monarchs, Ministers and Maps: The Emergence of Cartography as a Tool of Government in Early Modern Europe*. Chicago: University of Chicago Press, 1992.

Burke, Peter. *The Historical Anthropology of Early Modern Italy: Essays on Perception and Communication*. Cambridge: Cambridge University Press, 1987.

Burke, Peter, and Roy Porter, eds. *The Social History of Language*. Cambridge: Cambridge University Press, 1987.

Burroughs, Charles. "Spaces of Arbitration." In *Medieval Practices of Space*, edited by Barbara Hanawalt and Michal Kobialka. Minneapolis: University of Minnesota Press, 2000.

Busino, Giovanni. *Les théories de la bureaucratie*. Paris: Presses Universitaires de France, 1993.

Busquet, Raoul. *Histoire de Marseille*. Edited by Pierre Guiral. Paris: Éditions Robert Laffont, 1978.

——. "L'organisation de la justice à Marseille au moyen âge." *Provincia* 2 (1922): 1–15.

Cambridge Economic History, 2d ed. 3 vols. Edited by M. M. Postan and Edward Miller. Cambridge: Cambridge University Press, 1987.

Carruthers, Mary. *The Book of Memory: The Study of Memory in Medieval Culture.* Cambridge: Cambridge University Press, 1990.

Cassirer, Ernst. *The Myth of the State.* New Haven: Yale University Press, 1946.

Chambre de Commerce de Marseille. *Histoire du commerce de Marseille.* 7 vols. Edited by Gaston Rambert. Paris: Plon, 1949–1966.

Chiffoleau, Jacques. *La comptabilité de l'au-delà: les hommes, la mort et la religion dans la région d'Avignon au quatorzième siècle.* Rome: École Française de Rome, 1980.

Clanchy, M. T. *From Memory to Written Record: England, 1066–1307,* 2d ed. London: Blackwell, 1993.

Cohn, Bernard S., and Nicholas B. Dirks. "Beyond the Fringe: The Nation State, Colonialism, and the Technologies of Power." *Journal of Historical Sociology* 1 (1988): 224–29.

Cohn, Samuel Kline, Jr. *Death and Property in Siena, 1205–1800.* Baltimore: Johns Hopkins University Press, 1988.

——. *The Laboring Classes in Renaissance Florence.* New York: Academic Press, 1980.

Coulet, Noël. *Aix-en-Provence: éspace et relations d'une capitale (milieu XIV⁰ siècle–milieu XV⁰ siècle).* 2 vols. Aix-en-Provence: Université de Provence, 1988.

——. "Quartiers et communauté urbaine en Provence (XIIIᶜ–XVᶜ siècles)." In *Villes, bonnes villes, cités et capitales. Études d'histoire urbaine (XIIᵉ–XVIIIᵉ siècle) offertes à Bernard Chevalier,* edited by Monique Bourin, 351–59. Tours: Université de Tours, 1989.

Couton, G., and H.-J. Martin. "Une source d'histoire sociale: le registre de l'état d'âmes." *Revue d'histoire économique et sociale* 45 (1967): 244–53.

Crozier, Michael. *Le phénomène bureaucratique.* Paris: Seuil, 1963.

Dauzat, A. *Dictionnaire étymologique des noms de familles et prénoms de France.* Paris: Larousse, 1951.

Davies, Wendy, and Paul Fouracre, eds. *The Settlement of Disputes in Early Medieval Europe.* Cambridge: Cambridge University Press, 1986.

Denecke, Dietrich, and Gareth Shaw, eds. *Urban Historical Geography: Recent Progress in Britain and Germany.* Cambridge: Cambridge University Press, 1988.

Dilke, O. A. W. *Greek and Roman Maps.* Ithaca: Cornell University Press, 1985.

Douglas, Mary. *How Institutions Think.* Syracuse: Syracuse University Press, 1986.

Drendel, John. "Notarial Practice in Rural Provence in the Early Fourteenth Century." In *Urban and Rural Communities in Medieval France: Provence and Languedoc, 1000–1500,* edited by Kathryn L. Reyerson and John Drendel, 209–35. Leiden: Brill, 1998.

Droguet, Alain. *Administration financière et système fiscal à Marseille dans la seconde moitié du XIVᵉ siècle.* Aix-en-Provence: Cahiers du Centre d'Études des Sociétés Méditerranéennes, 1983.

Duby, Georges. *The Chivalrous Society.* Translated by Cynthia Postan. Berkeley: University of California Press, 1977.

Du Cange, Charles Du Fresne. *Glossarium mediae et infimae latinitatis.* Paris, 1842.

Dupanloup, Marc. "La corporation des cuiratiers à Marseille dans la première moitié du XIVᵉ siècle." *Provence historique* 77 (1969): 189–213.

Eckstein, Nicholas A. *The District of the Green Dragon: Neighbourhood Life and Social Change in Renaissance Florence.* Florence: Leo S. Olschki Editore, 1995.

Edgerton, Samuel Y., Jr. *The Renaissance Rediscovery of Linear Perspective.* New York: Basic Books, 1975.

Eisenstein, Elizabeth. *The Printing Press as an Agent of Change: Communications and Cultural Transformations in Early Modern Europe.* 2 vols. Cambridge: Cambridge University Press, 1979.

Emery, Richard W. *The Jews of Perpignan in the Thirteenth Century: An Economic Study Based on Notarial Records.* New York: Columbia University Press, 1959.

——. "The Use of the Surname in the Study of Medieval Economic History." *Medievalia et Humanistica* o.s., 7 (1952): 43–50.

L'enquête sur les catégories: de Durkheim à Sacks. Paris: École des Hautes Études en Sciences Sociales, 1994.

Fabre, Augustin. *Notice historique sur les anciennes rues de Marseille, démolies en 1862 pour la création de la rue Impériale.* Marseille, 1862.

——. *Les rues de Marseille.* 5 vols. Marseille, 1867–69.

Fabre, Ghislaine, and Thierry Lochard. *Montpellier: la ville médiévale.* Paris: Imprimerie Nationale, 1992.

Fawtier, Robert. "Comment le roi de France, au début du XIVe siècle, pouvait-il se représenter son royaume?" In *Mélanges offerts à M. Paul-E. Martin par ses amis, ses collègues, ses élèves.* Vol. 40 of *Mémoires et documents publiés par la société d'histoire et d'archéologie de Genève,* 65–77. Geneva: La Société d'Histoire et d'Archéologie de Genève, 1961.

Fentress, James, and Chris Wickham. *Social Memory.* Oxford: Blackwell, 1992.

Février, P. A. *Le développement urbain en Provence de l'époque romaine à la fin du XIVe siècle.* Paris: Éditions E. de Boccard, 1964.

Foucault, Michel. *Madness and Civilization: A History of Insanity in the Age of Reason.* Translated by Richard Howard. New York: Vintage Books, 1965.

——. *Power/Knowledge: Selected Interviews and Other Writings, 1972–77.* Translated by Colin Gordon. Edited by Colin Gordon et al. New York: Pantheon Books, 1980.

Frangenberg, Thomas. "Chorographies of Florence: The Use of City Views and City Plans in the Sixteenth Century." *Imago Mundi* 46 (1994): 41–64.

Frugoni, Chiara. *A Distant City: Images of Urban Experience in the Medieval World.* Translated by William McCuaig. Princeton: Princeton University Press, 1991.

Garrioch, David. "House Names, Shop Signs and Social Organization in Western European Cities, 1500–1900." *Urban History* 21 (1994): 20–48.

Geary, Patrick J. *Phantoms of Remembrance: Memory and Oblivion at the End of the First Millennium.* Princeton: Princeton University Press, 1995.

Geremek, Bronislaw. *The Margins of Society in Late Medieval Paris.* Translated by Jean Birrell. Cambridge: Cambridge University Press, 1987.

Giglioli, Pier Paolo, ed. *Language and Social Context.* Harmondsworth: Penguin, 1971.

Ginzburg, Carlo. "Clues: Roots of a Evidential Paradigm." In idem, *Clues, Myth, and Historical Method*, translated by John and Anne C. Tedeschi, 96–125. Baltimore: Johns Hopkins University Press, 1989.

Godding, Philippe, ed. *Le notariat en roman pays de Brabant et l'enseignement du notariat à l'université catholique de Louvain.* Brussels: Archives Générales du Royaume, 1980.

Goitein, S. D. *A Mediterranean Society: The Jewish Community of the Arab World as Portrayed in the Documents of the Cairo Geniza.* 5 vols. Berkeley: University of California Press, 1967.

Goody, Jack. *The Interface between the Written and the Oral.* Cambridge: Cambridge University Press, 1987.

——. *The Logic of Writing and the Organization of Society.* Cambridge: Cambridge University Press, 1986.

Gouron, André. *La réglementation des métiers en Languedoc au moyen âge.* Geneva: Droz, 1958.

Guenée, Bernard. "La géographie administrative de la France à la fin du moyen âge: élections et bailliages." *Le moyen âge* 67 (1961): 293–323.

Gumperz, John J. "The Speech Community." In *Language and Social Context*, edited by Pier Paolo Giglioli, 219–31. Harmondsworth: Penguin, 1971.

Gumperz, John J., and Dell Hymes, eds. *Directions in Sociolinguistics: The Ethnography of Communication.* New York: Holt, Rinehart and Winston, 1972.

Hacking, Ian. "Making Up People." In *Reconstructing Individualism: Autonomy, Individuality, and the Self in Western Thought*, edited by Thomas C. Heller, Morton Sosna, and David W. Wellbery. Stanford: Stanford University Press, 1986.

Halbwachs, Maurice. *On Collective Memory.* Edited and translated by Lewis A. Coser. Chicago: University of Chicago Press, 1992.

Hallam, Elizabeth. *Capetian France, 987–1328.* London: Longman, 1980.

Hardwick, Julie. *The Practice of Patriarchy: Gender and the Politics of Household Authority in Early Modern France.* University Park: Pennsylvania State University Press, 1998.

Harley, J. B. "Maps, Knowledge and Power." In *The Iconography of Landscape: Essays on the Symbolic Representation, Design, and Use of Past Environments*, edited by Denis Cosgrove and Stephen Daniels, 277–12. Cambridge: Cambridge University Press, 1988.

——. "Silences and Secrecy: The Hidden Agenda of Cartography in Early Modern Europe." *Imago Mundi* 40 (1988): 57–76.

Harley, J. B., and David Woodward, eds. *The History of Cartography.* 2 vols. Chicago: University of Chicago Press, 1987.

Harvey, P. D. A. *The History of Topographical Maps: Symbols, Pictures and Surveys.* London: Thames and Hudson, 1980.

——. "Local and Regional Cartography in Medieval Europe." In *Cartography in Prehistoric, Ancient, and Medieval Europe and the Mediterranean.* Vol. 1 of *The History of Cartography*, edited by J. B. Harley and David Woodward, 464–501. Chicago: University of Chicago Press, 1987.

——. *Maps in Tudor England.* Chicago: University of Chicago Press, 1993.

Hébert, Michel. *Tarascon au XVᵉ siècle: histoire d'une communauté urbaine provençale.* Aix-en-Provence: Édisud, 1979.

Heers, Jacques. *Espaces publics, espaces privés dans la ville. Le liber terminorum de Bologne (1294)*. Paris: CNRS, 1984.

——. *Family Clans in the Middle Ages: A Study of Political and Social Structures in Urban Areas*. Translated by Barry Herbert. Amsterdam: North-Holland Publishing, 1977.

——. *La ville au moyen âge en Occident*. Paris: Fayard, 1990.

——, ed. *Fortifications, portes de villes, places publiques, dans le monde méditerranéen*. Paris: Presses de l'Université de Paris-Sorbonne, 1985.

Helgerson, Richard. "The Land Speaks: Cartography, Chorography, and Subversion in Renaissance England." *Representations* 16 (1986): 50–85.

Herlihy, David. *Medieval and Renaissance Pistoia: The Social History of an Italian Town, 1200–1430*. New Haven: Yale University Press, 1967.

——. *Medieval Households*. Cambridge, Mass.: Harvard University Press, 1985.

——. *Pisa in the Early Renaissance: A Study of Urban Growth*. New Haven: Yale University Press, 1958.

——. "Problems of Record Linkages in Tuscan Fiscal Records of the Fifteenth Century." In *Identifying People in the Past*, edited by E. A. Wrigley, 41–56. London: Edward Arnold, 1973.

Herlihy, David, and Christiane Klapisch-Zuber. *Les toscans et leurs familles: une étude du catasto florentin de 1427*. Paris: École des Hautes Études en Sciences Sociales, 1978.

Herzfeld, Michael T. *The Social Production of Indifference*. Chicago: University of Chicago Press, 1992.

Hobsbawn, Eric, and Terence Ranger, eds. *The Invention of Tradition*. Cambridge: Cambridge University Press, 1983.

Hsiao, Kung-Chuan. *Rural China: Imperial Control in the Nineteenth Century*. Seattle: University of Washington Press, 1960.

Hsü, Immanuel C. Y. *The Rise of Modern China*. New York: Oxford University Press, 1970.

Hsu, Mei-Ling. "An Inquiry into Early Chinese Atlases through the Ming Dynasty." In *Images of the World: The Atlas through History*, edited by John A. Wolter and Ronald E. Grim. Washington D.C.: McGraw-Hill, 1997.

Hughes, Diane Owen. "Toward Historical Ethnography: Notarial Records and Family History in the Middle Ages." *Historical Methods Newsletter* 7 (1974): 61–71.

Huizinga, Johan. *The Autumn of the Middle Ages*. Translated by Rodney J. Payton and Ulrich Mammitzsch. Chicago: University of Chicago Press, 1996.

Jones, A. H. M. *The Roman Economy: Studies in Ancient Economic and Administrative History*. Edited by P. A. Brunt. Totowa, N.J.: Rowman and Littlefield, 1974.

Jones, Norman L. "William Cecil and the Making of Economic Policy in the 1560s and early 1570s." In *Political Thought and the Tudor Commonwealth: Deep Structure, Discourse and Disguise*, edited by Paul A. Fideler and T. F. Mayer. London: Routledge, 1992.

Jordan, William Chester. *Women and Credit in Pre-Industrial and Developing Societies*. Philadelphia: University of Pennsylvania Press, 1993.

Kain, Roger J. P., and Elizabeth Baigent. *The Cadastral Map in the Service of the State*. Chicago: University of Chicago Press, 1992.

Keene, Derek. *Survey of Medieval Winchester*. 2 vols. Oxford: Clarendon Press, 1985.

Keene, Derek, and Vanessa Harding. *A Survey of Documentary Sources for Property Holding in London before the Great Fire.* London: London Record Society, 1985.

Kent, Dale V. *The Rise of the Medici: Faction in Florence, 1426–1434.* Oxford: Oxford University Press, 1978.

Kent, Dale V., and F. W. Kent. *Neighbours and Neighbourhood in Renaissance Florence: The District of the Red Lion in the Fifteenth Century.* Locust Valley, N.Y.: J. J. Augustin, 1982.

Kent, F. W. *Household and Lineage in Renaissance Florence: The Family Life of the Capponi, Ginori and Rucellai.* Princeton: Princeton University Press, 1977.

Kirshner, Julius. "*Civitas Sibi Faciat Civem*: Bartolus of Sassoferrato's Doctrine on the Making of a Citizen." *Speculum* 48 (1973): 694–713.

——. "A Consilium of Rosello dei Roselli on the Meaning of 'Florentinus,' 'de Florentia' and 'de populo.'" *Bulletin of Medieval Canon Law* n.s., 6 (1976): 87–91.

——, ed. *The Origins of the State in Italy, 1300–1600.* Chicago: University of Chicago Press, 1995.

Klapisch-Zuber, Christiane. *Women, Family, and Ritual in Renaissance Italy.* Translated by Lydia G. Cochrane. Chicago: University of Chicago Press, 1985.

Konvitz, Josef. *Cartography in France, 1660–1848: Science, Engineering, and Statecraft.* Chicago: University of Chicago Press, 1993.

Kuehn, Thomas J. *Emancipation in Late Medieval Florence.* New Brunswick, N.J.: Rutgers University Press, 1982.

——. *Law, Family, and Women: Toward a Legal Anthropology of Renaissance Italy.* Chicago: University of Chicago Press, 1991.

Labov, William. *Sociolinguistic Patterns.* Philadelphia: University of Pennsylvania Press, 1972.

Laffont, Jean L., ed. *Notaires, notariat et société sous l'ancien régime. Actes du colloque de Toulouse, 15 et 16 décembre 1989.* Toulouse: Presses Universitaires du Mirail, 1990.

Lakoff, George. *Women, Fire, and Dangerous Things: What Categories Reveal about the Mind.* Chicago: University of Chicago Press, 1987.

Lakoff, George, and Mark Johnson. *Metaphors We Live By.* Chicago: University of Chicago Press, 1980.

Lane, Frederick C. "The Enlargement of the Great Council of Venice." In *Florilegium Historiale: Essays Presented to Wallace K. Ferguson,* edited by J. G. Rowe and W. H. Stockdale, 237–74. Toronto: University of Toronto Press, 1971.

Latouche, Robert. "Étude sur le notariat dans le comté de Nice pendant le moyen âge." *Le moyen âge* 37 (1927): 129–69.

Lavedan, Pierre. *Histoire de l'urbanisme: renaissance et temps modernes,* 2d ed. Paris: Henri Laurens, 1959.

——. *Représentations des villes dans l'art du moyen âge.* Paris: Vanoest, 1954.

Leguay, Jean-Pierre. *La rue au moyen âge.* Rennes: Ouest-France, 1984.

Léonard, Émile-G. *Histoire de Jeanne Ière, reine de Naples, comtesse de Provence (1343–1382).* 3 vols. Monaco: Imprimerie de Monaco, 1932.

Lesage, Georges. *Marseille angevine: recherches sur son évolution administrative, économique et urbaine de la victoire de Charles d'Anjou à l'arrivée de Jeanne 1re (1264–1348)*. Paris: Boccard, 1950.

Lestocquoy, Jean. *Aux origines de la bourgeoisie: les villes de Flandre et d'Italie sous le gouvernement des patriciens (XIe–XVe siècles)*. Paris: Presses Universitaires de France, 1952.

Lopez, Robert S. *The Commercial Revolution of the Middle Ages, 950–1350*. Englewood Cliffs, N.J.: Prentice-Hall, 1971.

——. "Concerning Surnames and Places of Origin." *Medievalia et Humanistica* o.s., 8 (1954): 6–16.

Mabilly, Philippe. *Les villes de Marseille au moyen âge. Ville supérieure et ville de la prévôté*. Marseille: Astier, 1905.

Maire Vigueur, Jean-Claude, ed. *D'une ville à l'autre: structures matérielles et organisation de l'espace dans les villes européennes (XIIIe–XVIe siècle)*. Rome: École Française de Rome, 1989.

Malaussena, P. L. *La vie en Provence orientale aux XIVe et XVe siècles. Un exemple: Grasse à travers les actes notariés*. Paris: Librarie Générale du Droit et Jurisprudence, 1969.

Maps of Authority: Conflict in the Medieval and Early Modern Urban Landscape. Special number of the *Journal of Medieval and Early Modern Studies* 26 (1996).

Marseille et ses rois de Naples: la diagonale angevine, 1265–1382. Aix-en-Provence: Édisud, 1988.

Martines, Lauro. *Lawyers and Statecraft in Renaissance Florence*. Princeton: Princeton University Press, 1968.

Matthews, C. M. *English Surnames*. New York: Charles Scribner's Sons, 1966.

Maurel, Christian. "Le prince et la cité: Marseille et ses rois . . . de Naples (fin XIIIe–fin XIVe siècles)." In *Marseille et ses rois de Naples, la diagonale angevine, 1265–1382*, 91–98. Aix-en-Provence: Édisud, 1988.

——. "Structures familiales et solidarités lignagères à Marseille au XVe siècle: autour de l'ascension sociale des Forbin." *Annales ESC* 41 (1986): 657–81.

McKitterick, Rosamond. *The Carolingians and the Written Word*. Cambridge: Cambridge University Press, 1989.

Merton, Robert K., et al., eds. *Reader in Bureaucracy*. New York: Free Press, 1952.

Miller, Naomi. "Mapping the City: Ptolemy's *Geography* in the Renaissance." In *Envisioning the City: Six Studies in Urban Cartography*, edited by David Buisseret, 34–74. Chicago: University of Chicago Press, 1998.

Miller, William Ian. *Bloodtaking and Peacemaking: Feud, Law, and Society in Saga Iceland*. Chicago: University of Chicago Press, 1990.

Montorzi, Mario. *Fides in rem publicam: ambiguità e techniche del diritto comune*. Naples: Casa Editrice Joven, 1984.

Moore, R. I. *The Formation of a Persecuting Society: Power and Deviance in Western Europe, 950–1250*. New York: Basil Blackwell, 1987.

Mousnier, Roland. *The Institutions of France under the Absolute Monarchy, 1598–1789*. Translated by Brian Pearce. Chicago: University of Chicago Press, 1979.

Mutius, Hans-Georg von, ed. *Jüdische Urkundenformulare aus Marseille in babylonisch-aramäischer Sprache*. Vol. 50 of *Judentum und Umwelt*. Frankfurt: Peter Land, 1994.

Nicolet, Claude. *Space, Geography, and Politics in the Early Roman Empire.* Translated by Hélène Leclerc. Ann Arbor: University of Michigan Press, 1991.

Noble, Thomas F. X. "Literacy and the Papal Government in Late Antiquity and the Early Middle Ages." In *The Uses of Literacy in Early Mediaeval Europe,* edited by Rosamond McKitterick, 82–108. Cambridge: Cambridge University Press, 1990.

Noiriel, Gérard. *Le creuset français: histoire de l'immigration XIXᵉ–XXᵉ siècles.* Paris: Seuil, 1988.

———. *La tyrannie du nation: le droit d'asile en Europe 1793–1993.* Paris: Calmann-Lévy, 1991.

Nora, Pierre, ed. *Realms of Memory: Rethinking the French Past.* Translated by Arthur Goldhammer. New York: Columbia University Press, 1996.

Norman, E. Herbert. *Origins of the Modern Japanese State.* Edited by John W. Dower. New York: Pantheon Books, 1975.

Il notaio nella civiltà fiorentina: secoli XIII–XVI: mostra nella biblioteca medicea laurenziana, Firenze, 1 Ottobre–10 Novembre 1984. Florence: Vallecchi Editore, 1984.

Otis, Leah Lydia. *Prostitution in Medieval Society: The History of an Urban Institution in Languedoc.* Chicago: University of Chicago Press, 1985.

Pansier, Pierre. *Histoire de la langue provençale à Avignon du 12ᵉ au 19ᵉ siècle.* Geneva: Slatkine Reprints, 1974 [1924].

Pernoud, Régine. *Essai sur l'histoire du port de Marseille des origines à la fin du XIIIᵉᵐᵉ siècle.* Marseille: A. Ged, 1935.

Petrucci, Armando. "Pouvoir de l'écriture, pouvoir sur l'écriture dans la renaissance italienne." *Annales ESC* 43 (1988): 823–47.

———. *Writers and Readers in Medieval Italy: Studies in the History of Written Culture.* Edited and translated by Charles M. Radding. New Haven: Yale University Press, 1995.

Pick, Shlomo H. "The Jewish Communities of Provence before the Expulsion in 1306." Ph.D. diss., Bar-Ilan University, 1996.

Pinker, Steven. *How the Mind Works.* New York: W. W. Norton and Co., 1997.

Pinto, John A. "Origins and Development of the Ichnographic City Plan." *Society of Architectural Historians Journal* 35 (1976): 35–50.

Platt, Colin. *Medieval Southampton: The Port and Trading Community, A.D. 1000–1600.* London: Routledge and Kegan Paul, 1973.

Plesner, Johan. *L'émigration de la campagne à la ville libre de Florence au XIIIᵉ siècle.* Translated by F. Gleizal. Copenhagen: Gyldendal, 1934.

Poisson, Jean-Paul. *Études notariales.* Paris: Economica, 1996.

———. *Notaires et société: travaux d'histoire et de sociologie notariales.* 2 vols. Paris: Economica, 1985–1990.

Poos, L. R. *A Rural Society after the Black Death: Essex, 1350–1525.* Cambridge: Cambridge University Press, 1991.

Pringle, James Keith. "The Quiet Conflict: Landlord and Merchant in the Planning of Marseille, 1750–1820." Ph.D. diss., Johns Hopkins University, 1984.

Pryor, John. *Business Contracts of Medieval Provence. Selected "Notulae" from the Cartulary of Giraud Amalric of Marseilles, 1248.* Toronto: Pontifical Institute of Mediaeval Studies, 1981.

Raeff, Marc. *The Well-Ordered Police State: Social and Institutional Change through Law in the Germanies.* New Haven: Yale University Press, 1983.

Raftis, J. Ambrose. *Tenure and Mobility: Studies in the Social History of the Mediaeval English Village.* Toronto: Pontifical Institute of Mediaeval Studies, 1964.

Reyerson, Kathryn L. *Business, Banking and Finance in Medieval Montpellier.* Toronto: Pontifical Institute of Mediaeval Studies, 1985.

——. "Land, Houses and Real Estate Investment in Montpellier: A Study of the Notarial Property Transactions, 1293–1348." *Studies in Medieval and Renaissance History* 6 (1983): 39–112.

——. *Society, Law, and Trade in Medieval Montpellier.* Aldershot: Variorum, 1995.

Rice, Sally. "Prepositional Prototypes." In *The Construal of Space in Language and Thought*, edited by Martin Pütz and René Dirven. Berlin: Mouton de Gruyter, 1996.

Riesenberg, Peter. *Citizenship in the Western Tradition: Plato to Rousseau.* Chapel Hill: University of North Carolina Press, 1992.

Romaine, Suzanne. *Socio-Historical Linguistics: Its Status and Methodology.* Cambridge: Cambridge University Press, 1982.

Romano, Dennis. *Patricians and Popolani: The Social Foundations of the Venetian Renaissance State.* Baltimore: Johns Hopkins University Press, 1987.

Rossiaud, Jacques. *Medieval Prostitution.* Translated by Lydia G. Cochrane. New York: Basil Blackwell, 1988.

Ruffi, Antoine. *Histoire de la ville de Marseille.* Marseille, 1652.

Sacks, Harvey. *Lectures on Conversation.* 2 vols. Edited by Gail Jefferson. Oxford: Blackwell, 1992.

Sahlins, Peter. *Boundaries: The Making of France and Spain in the Pyrenees.* Berkeley: University of California Press, 1989.

Salatiele. *Ars notarie.* 2 vols. Edited by Gianfranco Orlandelli. Milan: Giuffrè, 1961.

Schmid, Karl. *Gebetsgedenken und adliges Selbstverständnis im Mittelalter.* Sigmaringen: Jan Thorbecke Verlag, 1983.

Schneider, Jean. "Problèmes d'histoire urbaine dans la France médiévale." In *Tendances, perspectives et méthodes de l'histoire médiévale.* Vol. 1 of *Actes du 100ᵉ congrès national des sociétés savantes. Paris, 1975. Section de philologie et d'histoire jusqu'à 1610*, 137–62. Paris: Bibliothèque Nationale, 1977.

Schulz, Juergen. "Jacop de' Barbari's View of Venice: Map Making, City Views, and Moralized Geography before the Year 1500." *Art Bulletin* 60 (1978): 425–74.

Scott, James C. *Seeing Like a State: How Certain Schemes to Improve the Human Condition Have Failed.* New Haven: Yale University Press, 1998.

Shatzmiller, Joseph R. *Shylock Reconsidered: Jews, Moneylending, and Medieval Society.* Berkeley: University of California Press, 1990.

Silberman, Bernard S. *Cages of Reason: The Rise of the Rational State in France, Japan, the United States, and Great Britain.* Chicago: University of Chicago Press, 1993.

Sivéry, Gérard. "La description du royaume de France par les conseillers de Philippe Auguste et par leurs successeurs." *Le moyen âge* 90 (1984): 65–85.

Skelton, R. A. *Maps: A Historical Survey of Their Study and Collecting.* Chicago: University of Chicago Press, 1972.

Smail, Daniel Lord. "Accommodating Plague in Medieval Marseille." *Continuity and Change* 11 (1996): 11–41.

——. "Los archivos de conocimiento y la cultura legal de la publicidad en la Marsella medieval." *Hispania: revista española de historia* 57 (1997): 1049–77.

——. "The General Taille of Marseille, 1360–1361: A Social and Demographic Study." *Provence historique* 49 (1999): 473–85.

——. "Mapping Networks and Knowledge in Medieval Marseille: Variations on a Theme of Mobility." Ph.D. diss., University of Michigan, 1994.

——. "Notaries, Courts, and the Legal Culture of Late Medieval Marseille." In *Urban and Rural Communities in Medieval France: Provence and Languedoc, 1000–1500,* edited by Kathryn L. Reyerson and John Drendel, 23–50. Leiden: Brill, 1998.

——. "Telling Tales in Angevin Courts." *French Historical Studies* 20 (1997): 183–215.

——. "The Two Synagogues of Medieval Marseille: Documentary Evidence." *Revue des études juives* 154 (1995): 115–24.

Spiegel, Gabrielle M. *Romancing the Past: The Rise of Vernacular Prose Historiography in Thirteenth-Century France.* Berkeley: University of California Press, 1993.

Spilner, Paula Lois. "*Ut Civitas Amplietur*": Studies in Florentine Urban Development, 1282–1400." Ph.D. diss., Columbia University, 1987.

Stock, Brian. *The Implications of Literacy: Written Language and Models of Interpretation in the Eleventh and Twelfth Centuries.* Princeton: Princeton University Press, 1983.

Stouff, Louis. "Arles à la fin du moyen âge. Paysage urbain et géographie sociale." In *Le paysage urbain au moyen âge,* 225–51. Lyon: Presses Universitaires de Lyon, 1981.

——. "La population d'Arles au XVe siècle: composition socio-professionnelle, immigration, repartition topographique." In *Habiter la ville, XVe–XXe siècles,* edited by Maurice Garden and Yves Lequin, 7–24. Lyon: Presses Universitaires de Lyon, 1985.

——. "Les registres des notaires d'Arles (début XIVe siècle–1460). Quelques problèmes posés par l'utilisation des archives notariales." *Provence historique* 100 (1975): 305–24.

Suleiman, Ezra N. *Politics, Power, and Bureaucracy in France: The Administrative Elite.* Princeton: Princeton University Press, 1974.

——. *Private Power and Centralization in France: The Notaires and the State.* Princeton: Princeton University Press, 1987.

Teissier, Octave. *Marseille au moyen âge. Institutions municipales, topographies, plan de restitution de la ville.* Marseille, 1891.

Thongchai Winichakul. *Siam Mapped: A History of the Geo-Body of a Nation.* Honolulu: University of Hawaii Press, 1994.

Tocqueville, Alexis de. *The Old Régime and the French Revolution.* Translated by Stuart Gilbert. Garden City, N.Y.: Doubleday, 1955.

Trudgill, Peter. *Sociolinguistics: An Introduction to Language and Society.* Rev. ed. Harmondsworth: Penguin, 1983.

Turnbull, David. "Cartography and Science in Early Modern Europe: Mapping the Construction of Knowledge Spaces." *Imago Mundi* 48 (1996): 5–24.

——. "Constructing Knowledge Spaces and Locating Sites of Resistance in the Modern Cartographic Transformation." In *Social Cartography: Mapping Ways of Seeing Social and Educational Change,* edited by Rolland G. Paulston, 53–79. New York: Garland, 1996.

Vandeloise, Claude. *Spatial Prepositions: A Case Study from French.* Translated by Anna R. K. Bosch. Chicago: University of Chicago Press, 1991.

Vogler, Bernard, ed. *Les actes notariés, source de l'histoire sociale, XVIe–XIXe, actes du colloque de Strasbourg, mai 1978.* Strasbourg: Istra, 1979.

Weissman, Ronald F. E. *Ritual Brotherhood in Renaissance Florence.* New York: Academic Press, 1982.

Wernham, Monique. *La communauté juive de Salon-de-Provence d'après les actes notariés.* Toronto: Pontifical Institute of Mediaeval Studies, 1987.

Williams, Ernest N. *The Ancien Régime in Europe: Government and Society in the Major States, 1648–1789.* New York: Harper and Row, 1970.

Williams, Glyn. *Sociolinguistics: A Sociological Critique.* London: Routledge, 1992.

Woodward, David, ed. *Five Centuries of Map Printing.* Chicago: University of Chicago Press, 1975.

Zarb, Mireille. *Histoire d'une autonomie communale: les privilèges de la ville de Marseille du Xe siècle à la Révolution.* Paris: Picard, 1961.

Zonabend, Françoise. "Le nom de personne." *L'homme: revue française d'anthropologie* 20 (1980): 7–23.

Index

Page numbers in boldface type refer to illustrations and tables.